Storing Clocked Programs Inside DNA

A Simplifying Framework for Nanocomputing

Synthesis Lectures on Computer Science

Storing Clocked Programs Inside DNA: A Simplifying Framework for Nanocomputing

Jessica P. Chang and Dennis E. Shasha

ISBN: 978-3-031-00669-2   paperback
ISBN: 978-3-031-01797-1   ebook

DOI: 10.1007/978-3-031-01797-1

A Publication in the Springer series
SYNTHESIS LECTURES ON COMPUTER SCIENCE

Lecture #3
Series ISSN
Synthesis Lectures on Computer Science
Print 1932-1228   Electronic 1932-1686

Storing Clocked Programs Inside DNA

A Simplifying Framework for Nanocomputing

Jessica P. Chang and Dennis E. Shasha

Courant Institute, New York University

SYNTHESIS LECTURES ON COMPUTER SCIENCE #3

ABSTRACT

In the history of modern computation, large mechanical calculators preceded computers. A person would sit there punching keys according to a procedure and a number would eventually appear. Once calculators became fast enough, it became obvious that the critical path was the punching rather than the calculation itself. That is what made the stored program concept vital to further progress. Once the instructions were stored in the machine, the entire computation could run at the speed of the machine.

This book shows how to do the same thing for DNA computing. Rather than asking a robot or a person to pour in specific strands at different times in order to cause a DNA computation to occur (by analogy to a person punching numbers and operations into a mechanical calculator), the DNA instructions are stored within the solution and guide the entire computation. We show how to store straight line programs, conditionals, loops, and a rudimentary form of subroutines.

To achieve this goal, the book proposes a complete language for describing the intrinsic topology of DNA complexes and nanomachines, along with the dynamics of such a system. We then describe dynamic behavior using a set of basic transitions, which operate on a small neighborhood within a complex in a well-defined way. These transitions can be formalized as purely syntactical functions of the string representations.

Building on that foundation, the book proposes a novel machine motif which constitutes an instruction stack, allowing for the clocked release of an arbitrary sequence of DNA instruction or data strands. The clock mechanism is built of special strands of DNA called "tick" and "tock." Each time a "tick" and "tock" enter a DNA solution, a strand is released from an instruction stack (by analogy to the way in which as a clock cycle in an electronic computer causes a new instruction to enter a processing unit). As long as there remain strands on the stack, the next cycle will release a new instruction strand. Regardless of the actual strand or component to be released at any particular clock step, the "tick" and "tock" fuel strands remain the same, thus shifting the burden of work away from the end user of a machine and easing operation. Pre-loaded stacks enable the concept of a stored program to be realized as a physical DNA mechanism.

A conceptual example is given of such a stack operating a walker device. The stack allows for a user to operate such a clocked walker by means of simple repetition of adding two fuel types, in contrast to the previous mechanism of adding a unique fuel—at least 12 different types of strands—for each step of the mechanism.

We demonstrate by a series of experiments conducted in Ned Seeman's lab that it is possible to "initialize" a clocked stored program DNA machine. We end the book with a discussion of the design features of a programming language for clocked DNA programming. There is a lot left to do.

KEYWORDS

DNA computing, parallel programming, clocks, stored program concept, programming language design

Contents

Acknowledgments

We would like to thank Michelle Lynn Hall for assistance rendering chemical diagrams and proofreading. We would also like to thank Ada Zhang for her knowledge of laboratory procedures and insights therein. These helped with our preliminary modeling of reaction speed. A substantial thanks goes to Christofer Hedbrandh for his programming expertise in implementing a software engine that computes stable states. We did all our experiments in Professor Ned Seeman's lab. Ned also provided many invaluable reality checks.

Readers of the manuscript included Pearl Chin, Dalibor Frtunick, Panta da Silva, and Sarah Smith, all of whom provided excellent comments. Paul Rothemund's scientific criticisms improved the manuscript significantly.

Our editor, Diane Cerra, has been continuously helpful in finding outstanding reviewers, expediting editorial practices, and organizing the production team. That team included the very diligent Dr. C.L. Tondo and a very careful copyeditor, Pat Lenhardt.

Thanks to our editor Brendan Curry and to W. W. Norton for allowing us to reuse some of the illustrations from the book *Natural Computing*.

This work has been partially supported by the U.S. National Science Foundation under grants IIS-0414763, DBI-0445666, DBI-0421604, N2010 IOB-0519985, N2010 DBI-0519984, DBI-0421604, and MCB-0209754. This support is greatly appreciated.

Jessica P. Chang and Dennis E. Shasha
March 2011

CHAPTER 1

Introduction

Imagine the world before electronic computers. If an enterprise needed to figure out, say, a scheduling problem, it would have to call an expert who might come trotting along with a calculator. The expert would be needed because he or she would know the procedures to use. Nowadays, the enterprise can simply download scheduling software, store the software inside an electronic computer and execute it.

The current world of DNA computing is much like the calculator era. Researchers develop a sophisticated collection of DNA operations (analogous to designing and installing logarithms and trigonometric functions inside calculators or slide rules). This is useful and vital work. But if an operation requires N different strands, then the inventor must ship N vials containing those different strands to a site that is going to perform that operation. Moreover, the strands must be introduced into the solution in a specific order, so it might help to send an expert along just in case.

This book shows a way to construct clocked DNA programs that can be shipped from one site to another, much in the same way that scheduling software can be shipped around the world. An inventor (DNA software developer) will be able to create a program consisting of millions of DNA "sculptures" and then send them to labs all over the world. The lab would have to pour two "clock" strands, called "tick" and "tock," in an alternating fashion to run the DNA program.

We begin by defining a language that can describe states of a set of strands of DNA. Our description is graph theoretic, except in Chapter 3 when it is topological. (The topological description captures the order of strands around each junction, whereas the graph-theoretic description does not, but that ordering is unnecessary for our clocked stored programs.) We then use the language to describe transitions to move from one state to another. This gives us the ability to recapitulate all DNA machines that we know of. Finally, we describe a clocking mechanism that allows us to store clocked programs inside DNA that can mimic any machine in the literature and many more.

1.1 A SHORT HISTORY OF DNA COMPUTING

At a certain level of abstraction, a single DNA strand is just a sequence of letters from the alphabet A, C, G, and T. Two single strands will "hybridize" to one another to form a duplex strand if each A in one single strand is paired with a T of the other single strand, and each C in one single strand is paired with a G of the other. This is called Watson-Crick pairing, in honor of the primary discoverers of the DNA double-helix. Watson-Crick pairing provides the foundation for 90% of DNA computing.

DNA nanotechnology started with the pioneering work of Ned Seeman who was then at the state university of New York at Stony Brook. As Seeman tells it [Shasha and Lazere, 2010], two colleagues, Leonard Lerman and his post-doc Bruce Robinson, asked him to look at a DNA structure called a *Holliday Junction* (Figure 1.1). A Holliday junction is a four-stranded structure that resembles a four-way road intersection. Whereas stable DNA in nature has a linear topology, Lerman and Robinson were asking Seeman for help in building a model to explain how the nonlinear Holliday junction appears and disappears.

After helping his colleagues, Seeman decided that a more interesting question was how to create structures that would remain non-linear. Contemplating the Escher painting *Depth*, which shows a multi-dimensional school of fish, each having six propeller-like fins, Seeman noted that not only were the fish topologically six-arm junctions, but also that the fish were arranged like the molecules in a molecular crystal. There was periodicity front to back, top to bottom, and left to right.

Inspired by this, he constructed a variety of stable non-linear DNA structures[Seeman, 1982][Zhang and Seeman, 1994], making use of Watson-Crick pairing. The branch migration structure is a fundamental such stable structure. (Figure 1.2). Later, he and his students built a scafolding, known as a double-crossover (Figure 1.3), upon which to attach other DNA. Seeman's success in creating intricate nano-scale sculptures led many other labs to extend his works (there are over 50 labs in the world working on DNA nano-technology as of this writing).

We, however, will focus on efforts to make nucleotides do general-purpose computing. Charles Bennet [Bennett, 1973] did the earliest work on models for computing using molecular biological building blocks. His idea was to use RNA to perform some operations of a Turing machine. This work was theoretical but suggestive. DNA computing as we know it requires the ability to make DNA contort in controllable ways, an achievement that had to wait until Seeman's work and the startling paper of Leonard Adleman[Adleman, 1994]

Adleman used DNA to solve the directed Hamiltonian Path problem among cities. The Hamiltonian Path problem is to construct a path among simulated cities such that the path visits every city exactly once. Adleman showed that this problem could be "solved" in DNA by pouring various single strands into a solution and then fishing out promising strands using magnetized iron balls coated with the complements of the DNA representing cities. Because of the algorithm's historical importance and the techniques it illustrates, we describe Adleman's approach in some detail.

To start the process, Adleman represented cities as single strands of DNA and legs between cities as other single strands. Each "leg strand" had a left half that could bind the DNA for some departure city X (via Watson-Crick base pairing) and a right half that could bind the DNA for an arrival city Y (Figure 1.4). (As we will see in the physical chemistry section, every strand has an order. In this description, left half means the first half of the strand.) When all city and leg strands were mixed together, the leg strands held the city strands together in chains of various lengths.

Adleman located double strands whose length indicated that they had gone through exactly N cities (by using a technique called gel electrophoresis that can measure the length of strands). There was still the possibility that some of those strands included the same cities several times while missing other cities. So Adleman passed his solution through a series of filters that used magnetized

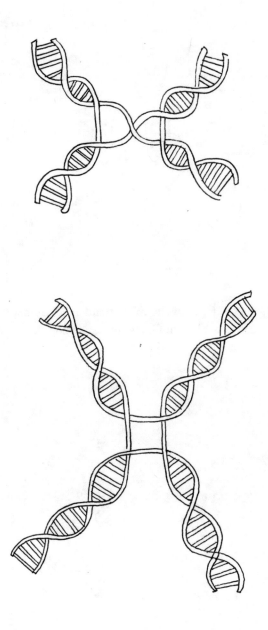

Figure 1.1: Two configurations of a four-way DNA intersection. In one configuration, the top right and the bottom left helices are linked. In the other, the top left and the top right are linked. The Watson-Crick attractions are equally long in the two cases, so either state can occur.

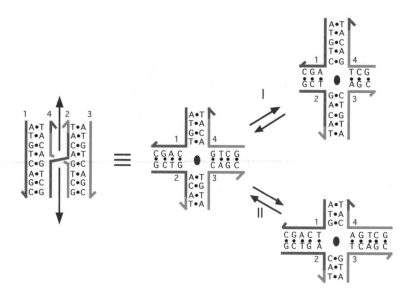

Figure 1.2: Branch Migration: Some of the ways in which strands of DNA may escape linearity. Watson-Crick pairing (A binds with T, and C binds with G) permits stable non-linear shapes (Courtesy of Ned Seeman.)

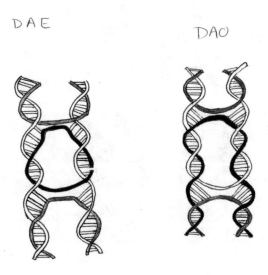

Figure 1.3: Double-crossover: This set of interlocking strands is quite strong at angstrom scales, thus providing a scaffold upon which to attach other DNA strands.

Figure 1.4: Adleman encoded each strand as a sequence of DNA bases (from the alphabet of A, C, T, G). For example, suppose for city B, the strand representing "To B" (*toB*) were ATT, and the symbol representing "From B" (*fromB*) were GTCT. Strand *toB'* would then be the complement of *toB* — TAA. Symbol *fromB'* would be CAGA. City B would be represented by the concatenation of *toB* and *fromB*. An edge from city Y to city B would be represented by the "leg" strand *fromY'* concatenated with *toB'*.

iron balls coated with DNA. Each ball was complementary to the DNA for some city. Any strands that stuck to all the balls were Hamiltonian paths. The idea that DNA could "solve" even a small instance of an NP-complete problem[1] captured the world's imagination.

Whereas Adleman had shown how to solve a particular problem, Paul Rothemund[Rothemund, 1996] showed that a Turing machine could be encoded using DNA and restriction enzymes. (He had, in fact, worked out most of these ideas before Adleman's paper appeared, but he couldn't convince his professors that his work was interesting. This demonstrates that even eminent professors can be as unimaginative as paperweights at times.) Rotheman's theoretical machine consisted of transitions encoded in DNA and restriction enzymes. When the transitions together with DNA ligase are applied to a state of the Turing machine (encoded as DNA), the machine enters a new state. Because a Turing Machine can simulate the behavior of any computer, this achievement showed that arbitrary computation was, in fact, possible.

Ogihara and Ray [Ogihara and Ray, 1999] propose a method for simulating Boolean circuits in DNA. Benenson et al. [Benenson, Paz-Elizur, Adar, Keinan, Livney, and Shapiro, 2001] again use restriction nuclease, DNA ligase, and ATP, as well as DNA that encodes both the "transition molecules" and the initial state of the computation. In their work, all inputs are mixed together and the chemical computation runs autonomously and asynchronously.

A very recent article by Qian et al.[Qian, Soloveichik and Winfree, 2011] has proposed a direct two stack implementation of a Turing machine. Transitions are designed as bi-directional, reusable "fuel" complexes built entirely from DNA. For example, a set of fuel complexes that effect the transition X Y → A B will transform a solution rich in strands X Y to one rich in strands A B.

[1]The Hamiltonian Path problem, like other NP-complete problem, has the property that the only guaranteed solution amounts to trying all possibilities. For the Hamiltonian Path problem, all possibilities means trying all possible paths.

The bi-directionality comes from the fact that the fuel complex is not destroyed by this transition. Instead, if there is an overwhelming concentration of A B later, the transition can produce X Y from A B. Using such transitions, Qian et al. show how to append onto and remove DNA strands from DNA stacks. The authors show that this is an efficient simulation of Turing machines. As in previous approaches, the computation runs autonomously and asynchronously. The trouble with asynchronous approaches is that each of these environments will have millions of Turing Machines. If they are asynchronous, then they can be in completely different and possibly interfering states.

As we will see in Chapter 4, Watson-Crick pairing leads to the construction of DNA walking robots and other nanomechanical devices. Such achievements are due to a remarkable observation by Bernard Yurke and his colleagues[Yurke, Turberfield, Mills, Simmel, and Neumann, 2000]: Watson-Crick pairing can be "competitive." That is, suppose single strand X is only partially paired with single strand Y. Simplifying slightly (by ignoring directionality), if X = AAAACCC and Y = AAAAGGG, then X and Y will pair at the right (where C hybridizes to G), not at the left. In their liquid romance, the parts of the strands that hybridize will embrace one another and the parts that don't will point away from one another. A tighter embrace leads to a more stable coupling. Suppose another strand Z (say AATTGGG) comes along that has a longer pairing with X (AAAACCC). Because this will lead to a tighter embrace, Z will "displace" Y using a zipper-like action as illustrated in the cartoon of Figure 1.5. Think California marriages.

Displacement will allow us to construct instruction stacks in a DNA solution. Each stack is a DNA sculpture containing heterogeneous strands wrapped together with fixed strands. Later, displacement will enable a simple protocol of pouring in other fixed strands over specific time intervals to peel away one strand (an "instruction") from each stack at a time. This leads to the possibility of parallel synchronous execution.

Martyn Amos [Amos, 2006] has written a friendly history of computing on biological substrates. He also offers well-founded speculations about the near and far future.

The goal of this monograph is much more specialized. We propose a way to encode arbitrary DNA operations in the natural model of a clocked parallel machine. In operation, this machine would store a program in DNA, which would later "execute" thanks to a steady drumbeat of tick and tock single strands. Just as clocking simplifies the design of electronic computers and effectively makes programming possible, our simple-to-build proposal supports branching, loops, and even subroutines. This will make it possible to ship DNA programs all over the planet and to run them using two test tubes of clock strands and perhaps a few magnetized iron balls.

1.2   UNDERLYING PHYSICAL CHEMISTRY

We lay out some basic features of the physical chemistry of DNA[Bath and Turberfield, 2007]. Understanding these features is not strictly necessary for the reader seeking to understand only the mathematical structure of DNA computing, but it provides physical intuition.

The DNA polymer (a polymer is just a long repeating chain of atom clusters) is composed of a flexible backbone of alternating sugar and phosphate residues. Each residue consists of a single

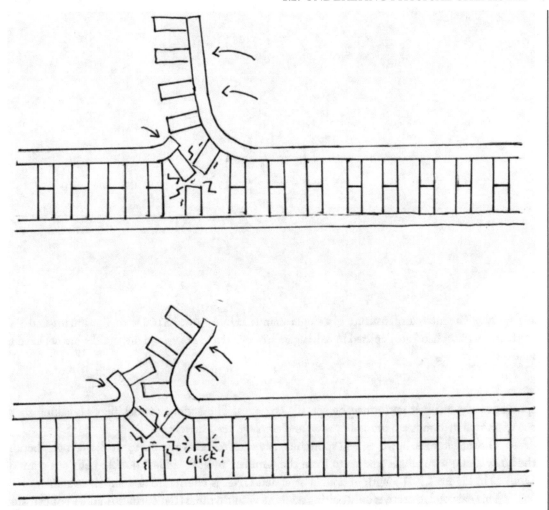

Figure 1.5: In the cartoon, the top right strand displaces the top left strand because the top right strand has a longer base pairing with the bottom strand.

organic base—adenine, thymine, cytosine, or guanine—bonded to a sugar, as shown in Figure 1.6. A chain of nucleotides connected along a single backbone is referred to as a *single* strand.

Because phosphates attach to the sugar ring at the 5' carbon when viewing the sequence of sugars in one direction and to the 3' carbon in the other direction, a strand has an *orientation*. We follow the convention of describing single strands from the 5' to 3' direction. For example, GATTACA denotes the bases, listed from the 5' end to the 3' end. Each phosphate-sugar-base unit is called a *nucleotide* or *base*. The binding and unbinding of these backbone linkages does not occur

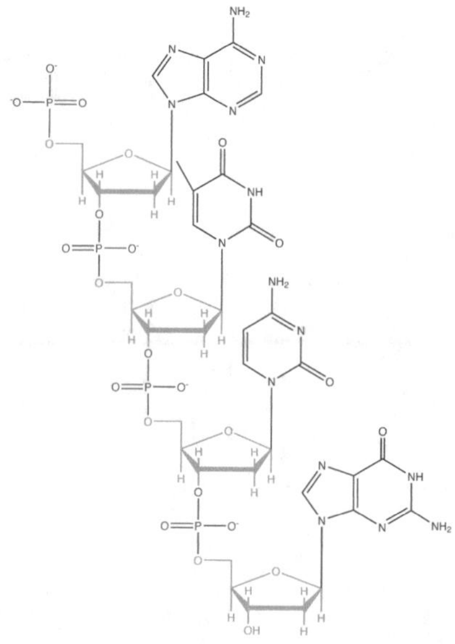

Figure 1.6: The molecular structure of a single stranded DNA, coded ATCG with 5' at the top and 3' at the bottom. Phosphates are depicted in red, sugars are depicted in green, and nucleotides are depicted in blue.

spontaneously, but it is performed by specific enzymes. The actions of making or breaking these sugar-phosphate bonds are termed *ligation* and *cleavage*, respectively.

The nucleotides, however, will spontaneously stick together thanks to weak hydrogen bonds, thus *hybridizing* two single strands to form the familiar "twisted ladder" double-helix structure of DNA as in Figure 1.7. A length of double-stranded DNA is referred to as a *duplex* strand.

This pairing is selective: only pairing adenine with thymine (the purines A and T) or cytosine with guanine (the pyrimadines C and G); Figure 1.8 illustrates this as interlocking notches. As mentioned above, this is called the Watson-Crick model of base-pair bonding. Due to the structure of the DNA polymer, two single strands will hybridize only if they are *anti-parallel*—their orientations are opposite one another—and will impart a right-handed twist to the duplex, and shown in Figure 1.8.

Because both the base-pairing and the backbone orientation are fixed for any given single strand, there is one and only one *complementary* sequence of nucleotides with which it will hybridize completely; this complement is obtained by swapping each base with its partner and reversing the orientation, thus GATTACA has the complement TGTAATC. By appropriately coding a collection of single strands, we can thus precisely control the manner in which they will hybridize.

The above is a slight simplification. There are, in fact, some exceptions and pathological cases that complicate the simple Watson-Crick model. Duplex DNA will change from a clockwise

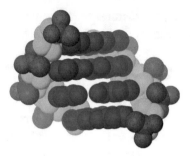

Figure 1.7: A 3-dimensional space-filling model of duplex DNA. The strand on the right is coded ATCG (oriented 5'–3' from top to bottom), and the strand on the left is coded CGAT (also oriented 5'–3' from bottom to top).

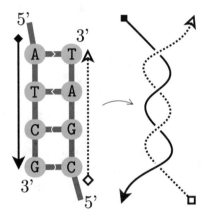

Figure 1.8: A cartoon demonstrating Watson-Crick pairing of bases by A–T and C–G hydrogen bonds. On the right is shown the right-handed twisting and anti-parallel backbone orientation.

helix to a counter-clockwise helix called Z-DNA under the right conditions. A long sequence of guanine bases will form a quadruplex helix, in which backbones alternate orientation. Triple-base bonding, partial bonds, "incorrect" pairings such as adenine with guanine, and more complicated configurations involving other molecules besides DNA are possible. For our purposes, we shall assume all binding follows strict Watson-Crick base pairing and forms the standard anti-parallel, right-handed duplex between a single strand and its complement. We make this assumption because exceptional cases can typically be avoided without too much difficulty.

The Watson-Crick model allows the construction of linear and branched DNA structures of quite sophisticated topology. A *complex* is the general term for such a structure, possibly involving many strands. For our purposes, such structures together with various rules and lab protocols for manipulating them may also be thought of as *machines*.

1.3   DNA COMPUTATION AND ENGINEERING

The relative chemical stability, specificity of binding, ease of synthesis, and size of the DNA molecule make DNA an ideal substrate for molecular computation. A particular strength of DNA computation is its inherent parallel nature. It is easy to replicate strands hundreds, or even billions of times, so many thousands of computational instructions can be carried out simultaneously.

This property suggests that DNA computation could be used to solve problems and perform computations that may otherwise scale poorly with conventional computational models, particularly certain types of NP-complete search algorithms, such as the traveling salesman problem[Shin, Zhang, and Jun, 1999]. Unfortunately, the combinatorial explosion of possibilities within NP-complete problems defeats DNA computing just as it defeats electronic computing. For example, finding the directed Hamiltonian path for 1,000 cities would require more nucleotides than there are atoms in the known universe.

Certain approaches to DNA computation, such as tile-based constructive computation[Winfree, 1996] [Winfree, Yang, and Seeman, 1998] [Cook, Rothemund and Winfree, 2004] have even been shown to be Turing complete. While various *direct* implementations of a Turing machine have been proposed—utilizing a physical component moving along a modifiable track[Rothemund, 1996]—none have yet been successfully implemented.

Parallelism, combined with the specificity of binding and structural properties, makes DNA ideal for nanoscale engineering and self-assembly, even at large scale. A pioneering example is Rothemund's recipe for DNA origami, a method that allowed for the creation of arbitrary 2-dimensional shapes out of (i) a single long (7,000+ base) strand of DNA from a virus and (ii) many short "staple" strands[Rothemund, 2006]. Whereas the single long virus strand is rather floppy when left on its own, a few hundred staple strands can, by Watson-Crick pairing, cause remote parts of the virus strand to stick together. Using staples, Rothemund was able to create a nanometer smiley face. Now there are software packages available for rapid prototyping of arbitrary 3D DNA nanostructure shapes assembled using this technique[Douglas, Marblestone, Teerapittayanon, Vazquez, Church and Shih, 2009].

Other researchers have built mobile structures with three specifically controllable conformations[Chakraborty, Sha, and Seeman, 2008], or walkers able to move payloads[Shin and Pierce, 2004]. These combined qualities make DNA a rich medium for nanorobotics, melding both computation and structural engineering.

In the next chapter, we suggest a language for describing complexes. Then we will describe transitions among complexes in terms of that language and eventually we use the transitions to show how to store programs inside DNA. Finally, we discuss the kinds of programming language that will arise when we can store clocked programs inside DNA.

CHAPTER 2

Notation

2.1 STATE DESCRIPTION

2.1.1 LABELING

Rather than explicitly listing every base and all base-pairings, we abstract the notation to use *labels* to refer to a particular sequence of bases (in the 5' to 3' direction). A single letter a may then be used to stand for the entire sequence of a single strand, such as GATTACA. The complement of such a sequence can then be denoted by a horizontal hat. For example, $\bar{a}$ would denote the sequence TGTAATC (again in the 5' to 3' direction). Square brackets will be used to demarcate a complex, so that the very simple example of a single strand of sequence a, with un-ligated 5' and 3' ends, would then be represented as $[a]$.

Figure 2.1: $[a]$ The smallest non-empty complex consists of just a single letter.

If there are multiple labels—a is GCTT and b is ATAC—then concatenation denotes their ligation. Thus, $[ab\bar{a}]$ is a single strand with the sequence GCTTATACAAGC; note that $\bar{a}$ is AAGC.

Figure 2.2: A complex demonstrating ligation $[ab\bar{a}]$.

Although we do not use such structures, if a single strand is ligated into a loop, this can be specified by enclosing a strand in parenthesis $[(ab\bar{a})]$, denoting that the 5' end hybridizes to the 3' end. This implies that the labels can be "rotated" and specify the same complex as in Figure 2.3.

Because there may be many copies of a label in a single machine, a *segment* refers to a particular instance of a labeled single strand. Each segment is identified by an index. For example, in the complex $[a_0ba_1]$, segment a_0 refers to the segment at the 5' end of the strand, while a_1 refers to the segment at the 3' end of the strand. Segments with different labels may share the same index, as in $[a_0b_0a_1]$ Omitting an index in the case of duplicate labels leads to ambiguity and so is considered a syntax error, as in $[xyx]$. Likewise, repeating the index of segments that share the same label is considered a syntax error, as in $[x_0yx_0]$.

Figure 2.3: The loop strand $[(ab\bar{a})] = [(b\bar{a}a)] = [(\bar{a}ab)]$.

For brevity, we sometimes use abstraction. The simplest abstraction uses standard ellipses to denote elision, e.g., $[a_0b_0a_1b_1 \ldots a_{99}b_{99}a_{100}b_{100}]$. Ellipses may also be used to elide irrelevant structure, such as focusing on two segments amidst a larger complex $[\ldots a\bar{a} \ldots]$. If this is the case, then we can substitute a variable or expression, such as $[a_0b_0a_1b_1 \ldots a_ib_ia_{i+1}b_{i+1} \ldots a_{100}b_{100}]$.

By choosing appropriate sequences, a designer can reduce extraneous complement symmetries among sequences to something like two or three base pairs. Because the hybridizing affinity of long matches vastly outweighs the hybridizing affinity of short matches, matches below a certain threshold length are ignored. We assume that either two segments will hybridize along their entire length in the case that their labels are complementary, or two segments will effectively have zero base-base interaction if their labels are not complementary. A machine for which this property holds true for every pair of labels is said to have a *consistent* labeling.

Labels can be split or merged in order to preserve consistency. For example, consider one machine containing the label a and another machine containing the label b, each with a consistent labeling when considered alone. Suppose the first half of a serves as the complement to the first half of b. In this case, we split the labels yielding $a = ec$ and $b = \bar{e}d$. Then we replace all occurrences of a and b, appropriately. Likewise, if we find that two labels, say g and h, are always ligated as $\ldots gh \ldots$ and that the same holds true for their complements as $\ldots \overline{hg} \ldots$, then the two labels can be merged, and we can say that $k = gh$. Note that due to the anti-parallel nature of hybridizing, the complement of concatenated labels are the commutated complements, i.e., $\bar{k} = \overline{gh} = \overline{h}\overline{g}$.

2.1.2   HYBRIDIZATION

When one segment is hybridized to another, each will indicate its partner through a superscript. So, for example, a complex referred to as a *hairpin* could be represented as $[\bar{a}_0^{a_0}b_0a_0^{\bar{a}_0}t_0]$. Because only complementary strands are able to hybridize, it is unnecessary to repeat the label of the hybridized partner, so simply superscripting the partner's index will suffice $[\bar{a}_0^0b_0a_0^0t_0]$. If the hybridized partner does not have an index—in the case where it is the only instance of that particular label—then an asterisk will serve as a placeholder, thus $[\bar{a}^*ba^*t]$ is equivalent to the hairpin above.

We use a vertical bar or a pipe character to demarcate different strands. A simple example of a stretch of duplex DNA consisting of two complementary strands hybridized together, as in Figure 2.5, would be $[a_0^0|\bar{a}_0^0]$ or $[a^*|\bar{a}^*]$ using the shorthand. The vertical bar thus implies a plurality of strands. In the case of Figure 2.4, we have a single unbroken strand, and thus the vertical bar notation does not apply. By contrast, because the complex in Figure 2.5 contains two distinct strands,

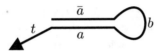

Figure 2.4: The hairpin $[\overline{a}^*ba^*t]$.

we need a vertical bar to demarcate the strand boundaries. Note that because the vertical bars are merely used as delimiters, the particular order in which the strands are listed is arbitrary: e.g., $[a^*|\overline{a}^*] = [\overline{a}^*|a^*]$.

Figure 2.5: A complex consisting of multiple non-continuous strands, such as $[a^*|\overline{a}^*]$, demonstrates the use of a vertical bar to demarcate separate strands. Note that the order in which the strands are listed is immaterial. So, the notation $[\overline{a}^*|a^*]$ also describes the complex above.

Using this notation, we can enumerate the various syntax errors which do *not* constitute a valid strand. In the following examples, the superscript of x_0 is invalid:

- no complementary segment is indexed by the superscript, such as in $[x_0^*]$ or $[x_0^1\overline{x}_0]$

- multiple complementary segments are indexed by the superscript, such as in $[\overline{x}_0x_0^*\overline{x}_1]$

- the complementary segment indexed by the superscript does not itself have a superscript $[x_0^*\overline{x}]$

- the complementary segment indexed by the superscript $[x_0^*\overline{x}_0^1x_1^0]$.

Note that, as currently specified, this notation allows for multiple disjoint and non-connected components to be listed as a single complex. For example, $[a^*b\overline{a}^*t|e^*|\overline{e}^*]$ is a syntactically legal construction. However, it consists of two entirely separable components, namely the hairpin $[a^*b\overline{a}^*t]$ and the duplex $[e^*|\overline{e}^*]$. Thus, a valid *component* is defined as a collection of strands that cannot be further decomposed into a legal construction without cutting strands or disrupting hybridizations. A *complex*, in general, denotes any set of valid components, and it can be listed using standard set notation. For example, we might say $\Omega = \{[a^*b\overline{a}^*t], [e^*|\overline{e}^*]\}$, where Ω denotes a complex consisting of multiple components. When representing a simple case such as one or two components and a single complex, we use a simple comma-separated list.

For the purpose of processing by computer, we specify a plaintext (ASCII-friendly) version of this notation, in which we avoid all use of bars and subscripts. In this instance, labels are always alphabetic characters, and complements are denoted by changing case, so $[ab\overline{a}]$ would become [a b A]. Segment indices are simply appended, rather than subscripted, so $[x_0yx_1]$ would become [x0 y x1]. In the case of a variable or expression, the index is enclosed in curly braces, so that

$[\ldots x_i \ldots]$ would become $[\ldots$ `x{i}` $\ldots]$ If there are more labels than alphabetic characters, labels can be multiple characters long. Because this can sometimes lead to confusion, one must separate labels with whitespace. For example, $[\alpha\beta\overline{\alpha}]$ might become `[alpha beta ALPHA]`. In order to denote hybridization without support for subscripts, one simply appends a caret and then the index of the hybridized partner, so the hairpin $[a^*b\overline{a}^*t]$ would become `[a^* b A^* t]` or `[a0^0 b0 A0^0 t0]`, depending on the level of verbosity desired.

Table 2.1: A summary of the notation syntax

notation	ascii	description
a	`a`	a label
$\overline{a}$	`A`	the complement of label a
$\overline{a}_2$	`A2`	segment indices distinguish multiple label instances
a^2	`a^2`	a =a is hybridized to the segment $\overline{a}_2$ =A2
$\overline{a}_2^*$	`A2^*`	$\overline{a}_2$ =A2 is hybridized to the only instance of a =a
a_{i+1}	`a{i+1}`	abstraction of numerals by arbitrary expressions
$\ldots$	$\ldots$	arbitrary segments and strands, or continuation
$\ldots ab \ldots$	`... a b ...`	the 3' end of a is ligated to the 5' end of b
$a\ldots\lvert\ldots b$	`a ... | ... b`	a and b are on different strands
$(a \ldots b)$	`(a ... b)`	the 3' end of b is ligated to the 5' end of a to form a loop
$[a^2\overline{a}_2^*]$	`[a^2 A2^*]`	brackets enclose a complete component or complex

2.1.3   NODES

You can think of the overall structure of the DNA complex as a formal graph structure consisting of nodes and edges. Edges will consist of either a directed single segment or an undirected duplex pair of hybridized segments. Nodes will be junctions where such edges meet; they will necessarily come in distinct types, based on the 5'–3' connectivity of the segments involved in such edges. Note that the "ordering" of the edges around a node matters; for this reason, the graph structure of nodes and edges alone is not sufficient to capture the full topology of such a DNA structure, as will be discussed in Chapter 3. The graph characterization is enough for our clocked stored program DNA computer and for the other machines we discuss, however.

2.2   TRANSITIONS

Because we want to model dynamic machines, we will formulate changes or transitions in terms of the representation from the previous section. In doing so, we ignore small perturbations of a complex, as they do not affect the overall topology of the DNA structure. We consider only those changes that affect the hybridization or ligation of segments.

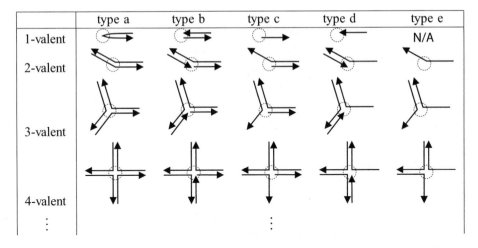

	type a	type b	type c	type d	type e
1-valent					N/A
2-valent					
3-valent					
4-valent					
⋮					

Figure 2.6: A chart of the various types of nodes. Each one represents a possible way in which strands can meet at a junction.

2.2.1 SPONTANEOUS TRANSITIONS

There are three basic transitions that will occur spontaneously in solution, provided the proper temperature and solvent environment is present.

2.2.1.1 Hybridization

Hybridization transitions characterize the simple hybridization of two unpaired single stranded segments when those segments are at least partly complementary. Consider the case where we have the segments a_i and $\overline{a}_j$. The transition $A_{a_i,\overline{a}_j}$ hybridizes these two segments to one another.

In our representation, this entails superscripting a_i with j to make a_i^j and transforming $\overline{a}_j$ into $\overline{a}_j^i$. Formally then, if we have un-hybridized segments a_i and $\overline{a}_j$ in some complex Ω_0, then we can define:

$$A_{a_i,\overline{a}_j} : \Omega_0 \to \Omega_1 \tag{2.1}$$

In the above, $A_{a_i,\overline{a}_j}(\Omega_0) = \Omega_1$ is the following mapping:

$$[\ldots a_i \ldots \overline{a}_j \ldots] \mapsto [\ldots a_i^j \ldots \overline{a}_j^i \ldots] \tag{2.2}$$

The two segments involved may be anywhere within the complex. Hybridization may involve segments in two different components, in which case these components are merged into a single component. For example, $A_{a,\overline{a}}([a], [\overline{a}]) = [a^*|\overline{a}^*]$.

2.2.1.2 Displace

Displace is called a branch migration. The *displace* transition replaces one single stranded segment with another of the same label already in an hybridized pair. Consider the single segment a_i along

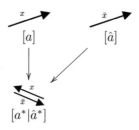

Figure 2.7: An example of the hybridize transition $A_{a,\bar{a}}$.

with the duplex pair a_j^k and $\bar{a}_k^j$. The transition D_{a_i,a_j} will then cause a_i to "displace" a_j, taking its place in the duplex pairing with $\bar{a}_k$ and leaving a_j un-hybridized. In the notation, this entails adding a superscript to a_i to yield a_i^k, removing the superscript from a_j^k to yield a_j and replacing the superscript of $\bar{a}_k^j$ to yield $\bar{a}_k^i$. Thus, if we have the segments (along with hybridizations) a_i, a_j^k, and $\bar{a}_k^j$ in some complex Ω_0, then we can define:

$$D_{a_i,a_j} : \Omega_0 \to \Omega_1 \tag{2.3}$$

In the above, $D_{a_i,a_j}(\Omega_0) = \Omega_1$ is the following mapping:

$$[\ldots a_i \ldots a_j^k \ldots \bar{a}_k^j \ldots] \mapsto [\ldots a_i^k \ldots a_j \ldots \bar{a}_k^i \ldots] \tag{2.4}$$

By convention, the segment that is initially un-hybridized is always listed as the first argument. Also, one needs to specify only two segments, as the third segment—in this case $\bar{a}_k$—can be determined from the subscript of the second argument.

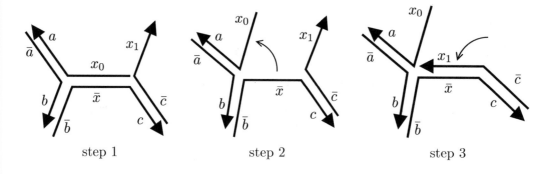

Figure 2.8: An example of the displace transition D_{a_1,a_0}. This is a non-separating displace and is thus reversible by applying D_{a_0,a_1} to the resulting complex.

A displace transition is *node-initiated* when the single segment must meet the duplex segments at some node. That node is called the *initiating node*; the node at the opposite end of the duplex segments is called the *terminating node*. When the terminating node is anything but type a (see Figure 2.6), the displace may cause the terminating node to be transformed into two distinct nodes. This is referred to as a *separating* transition.

In the non-separating case, the transition can be reversed by swapping arguments. That is, applying D_{a_i,a_j} first and then D_{a_j,a_i} immediately afterwards causes the complex to return to its original configuration. In the case of a separating transition, a displace transition sometimes results in two distinct components. For example, in the case of $D_{a_0,a_1}([a_0\bar{a}_1|a_1^*])$, the complex has a 2-valent initiating node of type c and a 1-valent terminating node of type b. This separating transition results in the disjoint components $[a_0^*\bar{a}^0]$ and $[a_1]$. The initiating node has become a 1-valent type a node, and the terminating node has separated to become a 1-valent node of type b and a 1-valent node of type c.

2.2.1.3 Exchange

This is sometimes called a duplex branch migration. The *exchange* transition takes two duplex segment pairs of the same label and exchanges their partners. Consider that we have the duplex pair a_i^j and $\bar{a}_j^i$ along with the duplex pair a_r^s and $\bar{a}_s^r$. The transition E_{a_i,a_r} will then cause a_i and a_r to effectively trade hybridizing partners with one another. The ligation does not change, however. This manifests in the notation as swapping the superscripts of a_i^j and a_r^s to yield a_i^s and a_r^j, and likewise swapping the superscripts of $\bar{a}_j^i$ and $\bar{a}_s^r$ to yield $\bar{a}_j^r$ and $\bar{a}_s^i$. For example, if we have a_i^j, $\bar{a}_j^i$, a_r^s, and $\bar{a}_s^r$ in some complex Ω_0, we define

$$E_{a_i,a_r} : \Omega_0 \to \Omega_1 \qquad (2.5)$$

where $E_{a_i,a_r}(\Omega_0) = \Omega_1$ is the mapping

$$[\ldots a_i^j \ldots a_r^s \ldots \bar{a}_j^i \ldots \bar{a}_s^r \ldots] \mapsto [\ldots a_i^s \ldots a_r^j \ldots \bar{a}_j^r \ldots \bar{a}_s^i \ldots] \qquad (2.6)$$

For the sake of convention and consistency with displace, the two segments specified should have identical labels. Also note the various symmetries of the operation. For a_i hybridized to $\bar{a}_j$ and a_r hybridized to $\bar{a}_s$, we have

$$E_{a_i,a_r} = E_{a_r,a_i} = E_{\bar{a}_j,\bar{a}_s} = E_{\bar{a}_s,\bar{a}_j} \qquad (2.7)$$

Like displace, an exchange transition is *node-initiated*. However, unlike displace, there are two terminating nodes. They are at the ends of the duplex segments opposite the initiating node. Whether or not an exchange is separating now depends on the types of both terminating nodes. If either is type a, then it is a non-separating transition. As with displace, the transition is reversible in the non-separating case. Any of its symmetrical forms serve as an inverse. So, applying E_{a_i,a_r} first and then E_{a_i,a_r} second results in no change to the complex. An example of the non-separating case is shown in Figure 2.9. An example of the separating case would be $E_{a_0,a_1}([a_0^0\bar{a}_1^1|a_1^1\bar{a}_0^0]) = [a_0^1\bar{a}_1^0], [a_1^0\bar{a}_0^1]$ in which two cross-hybridized strands separate into two separate continuous self-hybridized strands.

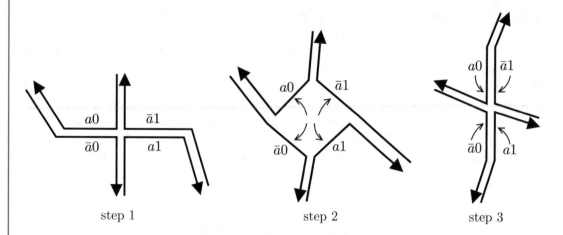

Figure 2.9: An example of the exchange transition $exchange_{a_1,a_0}$; this is a non-separating exchange and is thus reversible.

2.2.2   MANUAL TRANSITIONS

There also are several transitions that do not occur spontaneously but rather require either human intervention by way of lab protocol, the actions of proteins, or by other mechanisms.

2.2.2.1   Pour and Filter

These transitions are effectively inverses of one another. The pour transition consists of adding components to a complex to allow for interaction. An example of this might be $pour_{[b]}([a]) = [a], [b]$. In essence, pour is a simple multiset union.

Filter is the inverse of pour, removing specified components from a complex. So thus $filter_{[b]}([a], [b]) = [a]$. Thus, filter is a set partitioning.

In general, pour is straightforward to apply physically, as one can simply pour a vessel full of one complex into a vessel full of another complex. Filter, on the other hand, has no simple realization. One may filter on weight by use of a centrifuge, or by size by performing a gel-electrophoresis run, or by selecting on certain sequences by use of protein binding. Occasionally, these transitions do not correspond to actual laboratory operations or even physical changes, but they merely serve as a notational expedient. For example, certain "waste" components may no longer be able to interact in any transitions such as the component $[a_*|\bar{a}_*]$. Because such a duplex strand is stable, it does not affect further operation of the machine and therefore may be ignored.

2.2.2.2   Cleave and Ligate

These transitions correspond to joining or breaking covalent bonds along the DNA backbone. These are performed by proteins, such as DNA ligase, FokI, and others. These transitions manifest in the notation as the addition or removal of vertical pipes. For example, $cleave_{a_i,b_j}(\ldots a_i b_j) = \ldots a_i | b_j \ldots$. In the case of loop strands, the segments to be cleaved should first be rotated to be adjacent to the parenthesis, so that $|(\ldots a_i b_j \ldots)|$ should be first rewritten as $|(b_j \ldots a_i)|$. At this point, we have $cleave_{a_i,b_j} : |(b_j \ldots a_i)| \mapsto |b_j \ldots a_i|$. Remember that the arguments are always listed in a 5' to 3' order.

The ligate transition joins an open 3' endpoint and an open 5' endpoint. Notationally, this manifests as the simple deletion of a vertical pipe. Physically, this corresponds to the concatenation of two strands. Two segments can be ligated, provided one has an open 3' endpoint and the other has an open 5' endpoint; labeling is irrelevant. Strands should first be re-ordered to place the segments to be ligated adjacent to one another. So for example, $\ldots a_i | \ldots | b_j \ldots$ should become $\ldots a_i | b_j \ldots$ Thus, in general, we have $ligate_{a_i,b_j} : \ldots a_i | b_j \ldots \mapsto \ldots a_i b_j \ldots$

In the case that the segments to be ligated are the endpoints of the same strand, such as $|a_i \ldots b_j|$, then the application of ligate creates a loop strand and manifests as the addition of parentheses as $ligate_{a_i,b_j} : |a_i \ldots b_j| \mapsto |(a_i \ldots b_j)|$. As with cleave, the arguments are always listed in 5' to 3' order.

2.2.3   SYNTACTIC

Although not true transitions in the physical sense, these transitions allow for adjusting syntax without changing the underlying structure of the complex. In general, syntactic transitions may simply be performed in the notation. The following are all the syntactical transitions:

- Rotate the listing of segments in a loop strand.

- Rearrange the listing order of strands in a component.

- Add a numeral to the only occurrence of a label, and replace any asterisk that may refer to that segment with that numeral.

- Omit a numeral from the only occurrence of a label, then replace any subscripted numeral that may refer to that segment with an asterisk.

- Merge two labels as described in the Labeling subsection at the beginning of this chapter.

- Split two labels as described in the Labeling subsection.

- Rename all occurrences of a label to some unused label name.

2.3   DYNAMIC BEHAVIOR

With the basic transitions formalized, we can now describe the overall behavior of a DNA nanoma-chine system as a kind of state machine. Each state is represented by a particular complex—equivalently, a string or set of strings—and state transitions are drawn from the transitions described above. In general, because there may be multiple potential transitions for a given complex, the state transitions may constitute a non-deterministic state machine.

2.3.1   STABILIZATION

In solution, DNA machines naturally progress through available spontaneous transitions in a process we call *stabilization*. Not all transitions are equally favored, however; some are more likely to occur, or occur more quickly than others. Node-initiated transitions (hybridize, displace, and exchange) will occur more easily than non-node-initiated transitions. Among node-initiated transitions, hybridiza-tions occur rapidly and are energetically favorable. So, hybridizations have the highest precedence. Both displace and exchange transitions progress along segments from the initiating node in a ran-dom walk fashion, moving either forwards or backwards one base pair at a time. This makes them slower than a hybridization. Because, at each step, a displace must only break and make one base pair bond, whereas an exchange must make and break two, a displace transition will generally take prece-dence over an exchange. Except in cases of extremely high concentrations of DNA, intra-component transitions occur more frequently than inter-component transitions.

Thus, we can establish a rough (nested) precedence order for spontaneous transitions. Keep in mind that the deeper nested transitions are more similar in precedence to one another; in many cases, for example, it makes sense to treat all node-initiated transitions as having equal precedence.

1. intra-component transitions:

    (a) node-initiated transitions:

        i. hybridize

        ii. displace

        iii. exchange

    (b) disjoint hybridize

2. inter-component disjoint hybridize

With such a precedence established, the process of stabilization can be simulated by producing a *stabilization diagram*, which is a state machine diagram recording all component states and potential transitions. First, each component of the machine in question is recorded as if it were a distinct state. Then, each component in this set is examined to determine if any potential transitions (from high precedence to low) are available. For inter-component transitions, all components in the current iteration must be considered for possible interaction. The diagram is then expanded to include any new components that result from these new transitions, as well as adding in the transitions themselves.

This process is continued in successive stages until there are no longer any new states being introduced. Out of the resulting components, those that have no available transitions will be referred to as *stable*. There may also be cycles of states in which each state has an available transition, but the transitions stay within that set of states without preferring any one over the other. The states which make up such a cycle will be referred to as *semistable*. While the introduction of a precedence ordering diminishes the non-determinism to some extent, there can be cases which are inherently non-deterministic.

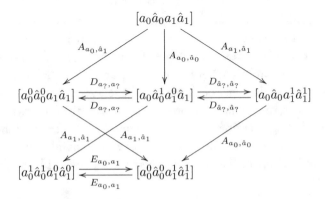

Figure 2.10: A state diagram for the stabilization of a particular strand, in this case $[a_0\overline{a}_0a_1\overline{a}_1]$. Note that the final two states are semi-stable. The hats in the figure are represented as carets.

2.3.2   STOICHIOMETRY

So far, we have ignored the particular concentrations of components. In many or perhaps even most cases, while there may be multiple available transitions at any given iteration, the transitions do not directly compete with one another. It does not matter in which order the transitions in question are performed because the resulting complex is the same. In effect, the transitions are commutative. By contrast, there are many cases in which transitions do compete, i.e., do not commute, and one set of transitions can lead to a different final state than another.

In those cases, concentrations can affect the operation of the machine. Thus, a state having a non-deterministic outcome when concentrations are unknown may turn into a state having a deterministic outcome when the stoichiometry—ratios between amount of transitioned components— is controlled. For example, for many transitions that involve multiple components—such as an inter-component hybridization—one of the components may serve as a limiting factor if the quantities are unequal. Determining the correct ratios between components requires a model.

Here is a naive model that is likely to be approximately correct in the large numbers world of a DNA solution. First, let Q_i represent the numerical quantity of the ith component. In the case of intra-component hybridizations, at a particular iteration of the stabilization process, suppose there are n_i distinct transitions involving the ith component. Then for a particular transition involving the ith component, *assume* the transition will occur on the quantity $\frac{Q_i}{n_i}$ of that component. For inter-component transitions, consider the combinatorial problem of drawing two components at random from a bag of quantities; this will be roughly the probability that they interact in solution. Thus, for a transition involving the ith component and the jth component, where there are a total of n_{ij} distinct transitions involving these two segments, *assume* the transition will be performed on the quantity $\frac{Q_i Q_j}{n_{ij} \sum_{\forall k} Q_k}$ taken from each of the two components.

This "perfect mixing" model may leave some components unchanged. At each iteration of this process, say ΔQ_i is the change in quantity of the ith component from the previous iteration to the current. If component i did not exist in the previous iteration, ΔQ_i will simply be the quantity of the component. Now we can introduce a threshold quantity ϵ in order to cut the tail off such asymptotes and make the stabilization calculation terminate. Any quantity below this threshold can effectively be ignored. Thus, the stabilization process will repeatedly progress through the transition precedence levels, distributing quantities in ratios determined by the formulas above, until $\Delta Q_i < \epsilon$ for all components. In collaboration with us, Christofer Hedbrandh has developed a stabilization algorithm based on this formula that takes stoichiometry into account. His software is implemented in python[Hedbrandh, 2010].

2.3.3   USER-LEVEL STATE DIAGRAM

Once the stabilized state is determined, the stabilization process can be thought of as a single step. Thus, we might think of stabilization as a *subroutine* and treat it as a single function applied to a complex. Subroutines may be used as state transitions to simplify the state diagram. A user can conceptualize the system as a state diagram from stabilized state to stabilized state. In such a state diagram, the transitions will consist of the combined subroutine of a manual transition—such as pour or filter—immediately followed by stabilize.

In practice, most DNA nanomachines or motifs consist of a single "main" component that progresses through several states. The dynamic behavior of this component is fueled by inputs that generally consist of simple strands or hairpins. As the main component progresses through these states, it produces "waste" components that don't interact further with the main component and therefore can be ignored.

CHAPTER 3

A Topological Description of DNA Computing

This chapter gives an alternate topological description of the statics and dynamics of our system. If you are a topologically-inclined thinker, then you might find this chapter very helpful. Otherwise, you can skip the chapter without loss of continuity.

3.1 AS A CELL COMPLEX

3.1.1 STATE

From a topological standpoint, we might choose to view a DNA complex as a cell/CW complex, or henceforth simply a *complex*. A complex captures the topology of an object by breaking it up into some finite number of *n-cells*—an *n*-dimensional object homeomorphic to an open ball—and recording the precise manner in which these cells are attached to one another. Additionally, we may also specify one of two *orientations* for a cell.

A 0-cell will correspond to the sugar-phosphate bond linking together nucleotides along the DNA backbone; we can view this as a 0-dimensional object, merely occurring at a single point. Orientation allows us to capture the 3' or 5' nature of these bonds. We can consider the point of attachment at the 5' end to be $(-)$ oriented and the point of attachment at the 3' end to be $(+)$ oriented.

A 1-cell then corresponds to a nucleotide, which we can view as a line segment connecting two 0-cells; the corresponding line segment will be oriented in the standard 5' to 3' direction. This allows us to capture the 5' / 3' nature of the DNA backbone by allowing two 1-cells to be attached to one another. The attachment consists of identifying or gluing together two oppositely oriented 0-cells into a single, un-oriented 0-cell. As a result, the orientation of the 1-cells will remain consistent across the attachment.

A 2-cell serves to represent the hydrogen bond between a pair of bases and manifests as a ribbon or a sheet between the two line segments; much as with the 1-cells, here the orientation is induced by the 1-cells, or nucleotides, running along either side. We might demonstrate this orientation in a diagram by coloring white the side from which it appears that the boundary 1-cells are oriented counter-clockwise, and coloring black the side from which it appears that the boundary 1-cells are oriented clockwise. The way in which 1-cells are attached determines whether the orientation coloring remains consistent along the length of duplex DNA. From a formal standpoint, we would also need two additional 1-cells "capping" the ribbon at either end, naturally extending to the process of "gluing" oppositely oriented capping 1-cells in order to attach 2-cells; we will come back to these later.

An *edge* will either be directed and consist of a single un-hybridized 1-cell, which we might refer to as a *simplex edge*, or it will be undirected and consist of a pair of hybridized 1-cells, which we might refer to as a *duplex edge*; in both cases, the edge is thought of as running parallel to the backbones.

A *node* will be a generalization of the Holliday junction, consisting of some combination of simplex and duplex edges meeting along a short line segment or loop. As duplex edges will connect to one another at a node via the "corners" of each ribbon-like 2-cell, the node will take roughly the form of a short loop or line segment with these "ribbons" hanging off. If we are looking down the "axis" of a node, the 2-cells will have a consistent orientation, such that the "white side" of every 2-cell faces clockwise, or every 2-cell faces counterclockwise. This is used to determine a consistent ordering in which to list the edges of a node. If we are to "flatten" the node out with edges pointing downwards and the "white side" of every 2-cell facing us, then the edges will be listed from left to right. In terms of the "cap" 1-cells mentioned previously, each node would be a contiguous chain of such cap 1-cells; the orientation of these cap 1-cells determines the ordering of the canonical list of edges.

3.1.2   TRANSITIONS

In a formal sense, each state is a homotopy equivalence class of embeddings. Transitions will be distinct topological events, corresponding to adding, deleting, separating, or joining together various cells. We will consider three basic types of transitions: Hybridize, Displace, and Exchange. These will be much like the Riedermeister moves of knot theory. However, unlike Riedermeister moves, there is a directionality to these transitions. They will operate on a 3-dimensional embedding, rather than a 2-dimensional projection. All transitions will consist of topological surgery, which replaces the interior of a small bounding sphere containing a specific arrangement of cells, with another specific arrangement of cells, while keeping the surface or "shell" of this bounding sphere fixed. In general, for the three basic transition types, the bounding sphere will contain some number of roughly parallel (or anti-parallel) 1-cells, with their bounding 0-cells lying on the surface of the sphere; these 1-cells remain more-or-less fixed throughout the transition, while various additions and deletions are made to the 2-cells within the bounding sphere. Attention must also be paid to correct label inverse behavior, and all hybridizations in both the start and end state must be "legal" in this sense.

The *Hybridize* transition represents a simple hybridization of unpaired nucleotides. The initial state of the bounding sphere will contain two 1-cells corresponding to an un-hybridized pair of matching nucleotides: one is oriented "upwards" which we might refer to as a, and one is oriented "downwards" which we might refer to as b. The end state of the bounding sphere will contain the same two 1-cells, now with the addition of a 2-cell between a and b. The transition is naturally unidirectional because hybridized base pairs typically do not spontaneously separate.

The *Displace* transition represents the displacement of one hybridized nucleotide by an unhybridized nucleotide. The initial state of the bounding sphere will contain three 1-cells with one oriented "upwards," which we might refer to as a, and two oriented "downwards," which we might refer to as b_1 and b_2. Initially, there will be a 2-cell between a and b_1. The end state is obtained by the deletion of this initial 2-cell and the addition of another between a and b_2. Because the initial state and the end state are related by symmetry, this transition will naturally be bidirectional. However, we might also form an equivalent formulation by treating the different "chiralities" of such a transition as distinct unidirectional transitions; this is analogous to the case of Reidermeister moves where we might treat the various symmetries as distinct from one another.

The *Exchange* transition represents the swapping of nucleotides between two hybridized pairs; this transition provides for the mobility of a classical Holliday junction. The initial state will have four 1-cells with two oriented "upwards," which we might refer to as a_1 and a_2, and two oriented "downwards," which we might refer to as b_1 and a_2. These orientations alternate as one travels (roughly) around the equator of the sphere. Listing them counterclockwise, we might have a_1, b_1, a_2, b_2. There will also be two 2-cells, one between a_1 and b_1 and another between a_2 and b_2. The end state is obtained by the deletion of these 2-cells and the addition of two other 2-cells, one between a_1 and b_2 and another between a_2 and b_1. Likewise as with Displace, the initial state and end state are related by symmetry, making the transition a bidirectional one, and we might also treat the various chiralities as distinct unidirectional transitions.

We might also consider transitions such as *Pour* and *Filter* that add or subtract components to the complex. Viewed as straightforward disjoint union and intersection operations, these correspond to pouring two solutions together or mechanically filtering out particulates and should be treated as unidirectional transitions.

CHAPTER 4

Machines and Motifs

In this chapter, we show how to build up a clocked programming system based on the transitions introduced in Chapter 2. We start by looking at larger scale transitions called motifs and then describe an instruction stack, loops, conditionals, and finally an example application. Because this builds substantially on earlier research, we begin with a brief review in which we describe previous construction techniques using our simple language of states and transitions.

4.1 A BRIEF REVIEW OF DNA CONSTRUCTION MOTIFS

4.1.1 TAPE MOTIF

Many machines use a motif that selectively attaches and detaches two segments end to end. The motif makes use of three labels, along with their complements. The segments to be attached to one another will be labeled a and b; the additional third label t will serve as a "tab" specific to the operation of the motif. While there may be several different segments labeled a and b, the motif attaches any un-hybridized instance of a to any un-hybridized instance of b arbitrarily. Although this non-deterministic behavior may suffice in some circumstances, it is usually better to constrain the un-hybridized instances of a and b so as to avoid ambiguity.

In addition to the a and b labels already present in the target machine, the motif also makes use of two additional strands, which are complements of one another. We might refer to these as $tape_{ab}$ and $untape_{ab}$.

In the example shown in Figure 4.3, we have $tape_{ab} = [\bar{t}\bar{b}\bar{a}]$, and thus $untape_{ab} = [abt]$. The strand $tape_{ab}$ will hybridize to both a and b, thus attaching them end to end and leaving an unhybridized tab $\bar{t}$. The strand $untape_{ab} = \overline{tape_{ab}}$ is the complement of $tape_{ab}$. Thus, by pouring in $untape_{ab}$, the tab is hybridized, and $tape_{ab}$ is displaced because the sequence of hybridized pairs is longer[Yurke, Turberfield, Mills, Simmel, and Neumann, 2000] [Zhang and Seelig, 2011].

This leaves a and b again unhybridized, along with an absolutely stable duplex component formed by the hybridized tape and untape strands. This stable duplex will not react with any other components, so it can safely be ignored, and we will consider it simply as *waste*. The stabilization process of both the untaping and the taping are demonstrated in Figure 4.1 and Figure 4.2, respectively.

Because these transitions are strongly unidirectional, we don't need to concern ourselves with the details of the stabilization diagram. A user can view the situation as the following: add various fuel strands as inputs and see the stabilized complexes as a result. To reflect this simplified viewpoint,

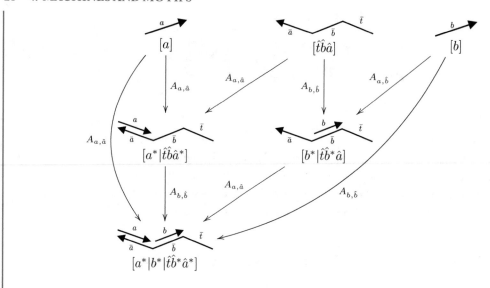

Figure 4.1: The complete stabilization diagram for the tape motif. The hats in the figure are represented as carets.

we introduce a state diagram for the machine in Figure 4.3. This will be a simple two state machine with inputs and outputs.

4.1.2 WALKERS AND GEARS

By using a repeated tape motif, we can selectively cause feet to "walk" alternately across a track of footholds. The tape motif can also cause the teeth of gears to mesh with one another. This concept has been used to develop walkers with a standard alternating bipedal gait [Shin and Pierce, 2004] and walkers having a staggered gait in which they keep one "foot" constantly forward[Sherman and Seeman, 2004]. In order to impose directionality, there must be at least three differently labeled footholds on the track, such that stepping forward with a leg would necessitate a different tape than stepping backwards. So consider the feet to be labeled by x and y and consider the track footholds to be labeled by a, b, and c.

Because each step requires a tape strand to place the foot and an untape strand to lift it, there will a total of 12 different fuel strands. Thus, the walker is effectively a cyclical 12-state machine. During each cycle, the walker progresses six steps along the track. Of course, this is only one among many variants.

4.1.3 POLYMER GROWTH

Another example of a machine motif is the growing polymer machine[Venkataraman, Dirks, Rothemund, Winfree, and Pierce, 2007]. The polymer consists

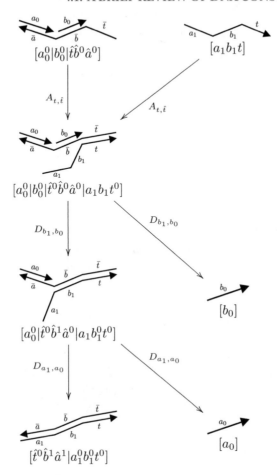

Figure 4.2: The complete stabilization diagram for the untape motif. Note that the displace transitions are separating, thus guaranteeing that the stabilization is unidirectional, and the resulting components are stable and not merely semistable.

of double stranded DNA, one of whose endpoints will be called an anchor and the other a payload. Both endpoints remain attached to the same unique polymer throughout the operation of the machine. Growth of the polymer occurs from the middle. This machine operates spontaneously and asynchronously in solution, growing as long as there is fuel in solution. Two types of hairpins are used to fuel the device. Each hairpin fuel will first hybridize onto the polymer, and then insert itself into the polymer by way of a double branch migration.

An application of this machine might be to separate two payloads. By combining several such mechanisms to a single pair of payloads, much larger payloads could be moved; however, at a certain distance, the stiffness of duplex DNA may be overwhelmed. By using appropriately modified fuels, it

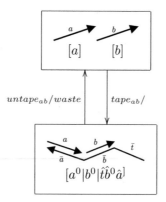

Figure 4.3: A simplified state diagram for the tape motif, in which the segments a and b are taped together. Transitions are listed with *input/output* notation. Note that $tape_{ab} = [\bar{t}^0\bar{b}^0\bar{a}^0]$ and $untape_{ab} = [abt]$. Also, in this case, we have $waste = [\bar{t}^0\bar{b}^0\bar{a}^0|a^0b^0t^0]$. Because the waste is hybridized, it is stable, so plays no further role in any reaction.

is also possible to leave single stranded sticky ends at regular intervals along the polymer. Thus, such a machine could be modified to grow tracks for walkers. For that purpose, more fuel types would be needed in order to provide the pattern of three distinct footholds necessary to impose directionality. Alternatively, when multiple polymers are used in parallel, such sticky ends might be used to anchor lateral support structures in the manner of a truss.

4.2   A CLOCKED STORED PROGRAM DNA MACHINE

The machines that have been described make use of a single primary component interacting with simple fuel strands or hairpins. For example, we might describe the earlier tape-untape machine as in Figure 4.3. This concept can be naturally extended to multiple state machines with interacting inputs (in the form of fuel strands) and outputs (waste strands). One can anchor such state machines to some easily manipulated substrate, such as 7 kilobase DNA origami[Rothemund, 2006]. The anchors would consist of "tabs" similar to those used in the walker motif and can be attached to the substrate via hybridization, ligation, or proteins with sequence specific binding sites. By releasing a tab, a user might remove an entire programmatic "module" if needed. This setup would thus allow for the removal or addition of large numbers of possibly heterogeneous components at will.

One intriguing application of such a module would consist of a stack motif that releases precisely programmable output patterns with support for flow control, looping, variable iteration, and more. This simple new mechanism allows clocked stored program DNA computing.

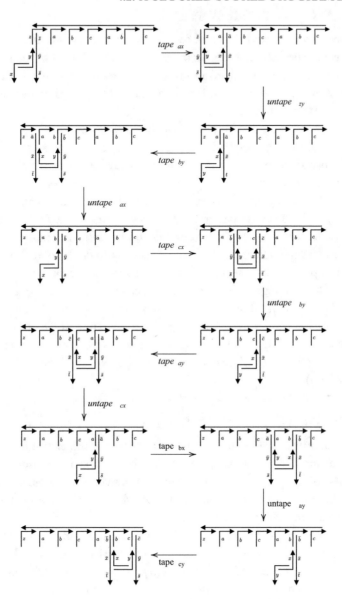

Figure 4.4: The cyclical walker motif. Only one direction is annotated, but by following the arrows backwards and substituting *tape* for the corresponding *untape* and vice-versa, the walker could easily move the other direction along the track. Waste produced by the *untape* steps is omitted for clarity, as waste strands play no further role. Note that the toehold label alternates to avoid unwanted interactions. Also, note that *untape*$_{bx}$ will essentially transition from the last state to the first—albeit with a 6-step shift—thus completing the cycle.

4.2.1   CLOCKED INSTRUCTION STACK

As we've seen, many machines require specific sequences of fuel strands or fuel hairpins to be added to the reaction vessel in a specific sequence; for our purposes, we will refer to such strands or hairpins as *instructions*. Here we propose a mechanism for storing a sequence of such instructions in a stack and then releasing them in a controlled way based on a clock. Our clocking mechanism requires two distinct fuel strands that we call *tick* and *tock*.

Assuming the stack is already built, it will operate as follows. First, *tick* is added to solution, which partially hybridizes to a corresponding exposed sticky end of the stack. The remaining end of the tick strand proceeds to displace the remainder of the strand to which it is hybridized by way of a single branch migration. This produces a waste component of duplex DNA, consisting of the tick strand and its complement, leaving an exposed sticky end on the stack. Second, *tock* is added to solution, which likewise partially hybridizes to this sticky end. Once again, a single branch migration displaces tock's complement and ligates to the instruction payload. This releases the instruction payload from the stack and ligates the payload to a duplex of tock and its complement.

In addition to operating as a sequence of instructions, the stack can operate as a standard stack data structure. The two stage cycle of *tick* and *tock* can be thought of as a single *pop* operation. Similarly, the two stage cycle of adding the desired instruction, along with an additional "taping" strand—namely, tick's complement—implements the *push* operation. Repeated pushing can be used to synthesize such stacks. (Pushing instructions onto data stacks is one classic way in which malicious hackers gain control of other people's machines. Hacking DNA machines in the same way could eventually be a problem.)

4.2.2   LINEAR (STRAIGHT LINE) PROGRAM

The actual mechanism operates in a similar manner to the tape motif, except that tabs are staggered and overlap with one another. For simplicity, we will refer to every instruction payload as a single segment p, regardless of hairpin topology or instruction individuality, because this does not affect the functioning of the instruction stack, provided payload segments have sequences that prevent them from binding inappropriately to tick or tock.

The basic alphabet required to make the clock work consists of two strands denoted by the labels α and β, along with their complements $\overline{\alpha}, \overline{\beta}$. Pairs of these basic units will then be concatenated to form four distinct connected strands: $[p\alpha\beta]$ and $[\overline{\alpha}\overline{\beta}]$ make up the stack component; $[\overline{\beta}\overline{\alpha}]$ and $[\beta\alpha]$ play the role of *tick* and *tock*, respectively. A stack component, consisting of three instructions, could then be constructed as shown in Figure 4.7.

The operation of the machine will be as follows. In the initial state of the machine, the un-hybridized $\overline{\alpha}$ segment acts as a toehold. The ligated $\overline{\beta}$ segment serves to block the release of the first instruction. The addition of the first *tick* strand ($[\overline{\beta}\overline{\alpha}]$) in the proper stoichiometry serves to displace this strand from the right and ready the first instruction. The addition of *tock* ($[\beta\alpha]$) then displaces and releases the first instruction from the stack, thus returning the stack to a state analogous to the initial state but with one less instruction on the right.

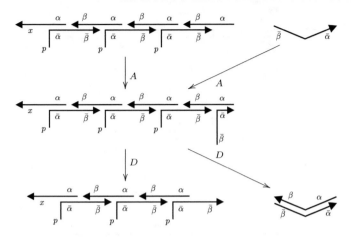

Figure 4.5: The stabilization diagram for the *tick* motif of the stack consists of a hybridization followed by a separating displace.

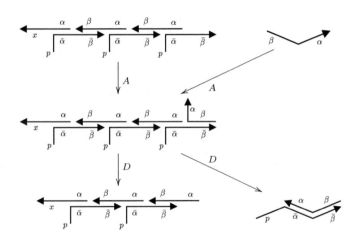

Figure 4.6: The stabilization diagram for the *tock* motif of the stack. Tock stabilization is essentially identical to the *tick* stabilization, except that the released "waste" double strand is linked to the desired instruction payload, here denoted by *p*.

This process can be repeated as long as there are instructions available on the stack. We can utilize the *x* segment in the "last" (leftmost) strand to anchor the stack to other components or protein structures with appropriate binding sites. Several variations of such motifs are possible. These will be discussed in Section 4.2.6.

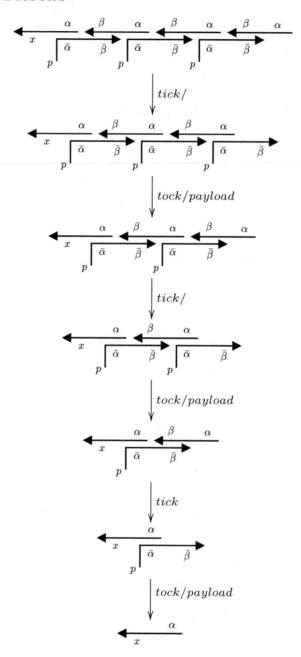

Figure 4.7: The operation of a stack from the user's perspective. This stack contains three instruction payloads, each of which consists of simply the label p. The label x may be used to anchor the stack to some substrate such as DNA origami or some protein with binding specificity.

While we can describe the state of a stack using the formalized component syntax, this quickly becomes cumbersome. For that reason, it is useful to use a more streamlined stack notation. So, for example, we might show the stack of Figure 4.7 as

> *tick/*
> *tock/*payload
> *tick/*
> *tock/*payload
> *tick/*
> *tock/*payload

4.2.3 GATING AND FLOW CONTROL

The previous example showed a straight line program. We can introduce branching behavior by using a strand-controlled gating mechanism. Attaching an initial *gate* strand $[\overline{\alpha\mu}]$ to the toehold in the initial state stops the stack from operating. However, upon addition of the strand $[\mu\alpha]$—which we might deem *unlock*$_\mu$—the gating strand will be displaced. Unlocking places the stack in a ready state, so subsequent additions of *tick* and *tock* release instruction strands as usual.

This motif allows a compiler to support branching and other program control functionality by building multiple stack groups, each having a different gate strand. Depending on the sequence of some particular unlocking μ, a stack from one group or another would be unlocked. For example, there may be one stack keyed to the label μ, another stack keyed to the label ν. Both still operate using the same clock strands once unlocked. This allows for zero, one, or both of these instruction blocks to be executed depending on the presence of zero, one, or both of $[\mu\alpha]$ and $[\nu\alpha]$. This permits a kind of trigger-based programming in which unknown inputs cause a DNA program to "branch" to one stack or another because the inputs unlock a particular strand.

The resulting program snippet begins with an unlock operation and then clocks through an arbitrary number of payload instructions as shown in the box below.

> *unlock*$_\mu$/
> *tick/*
> *tock/*payload
> *tick/*
> *tock/*payload
> *tick/*
> *tock/*payload

4.2.4 LOOPING

There is also the possibility of looping behavior if stoichiometry is appropriately set. Suppose there are n target machines and mn copies of the stack. By introducing n unlock strands, n stacks will be unlocked. Subsequent tick and tock strands will execute the instructions in those stacks. This leaves

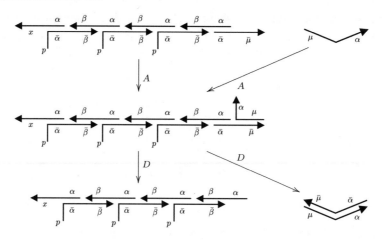

Figure 4.8: The stabilization diagram for a gated stack. First, there is a hybridization and then a displacement. Unlocking is thus similar to a *tock*.

$n(m-1)$ gated copies of the stack remaining. If each copy of the stack then includes a copy of the unlock strand as its last instruction, another n stacks will be unlocked after the first set of n stacks complete. That is, each execution of n stacks corresponds to an iteration of the loop.

The looping program snippet begins with an unlock operation and then clocks through an arbitrary number of payload instructions ending in an unlock operation, as shown in the box below.

$$unlock_\mu/$$
$$tick/$$
$$tock/\text{payload}$$
$$tick/$$
$$tock/\text{payload}$$
$$tick/$$
$$tock/\text{payload}$$
$$tick/$$
$$tock/unlock_\mu$$

4.2.5   ITERATED VARIABLES

By combining this technique with other gates, *for* and *while* statements are possible. For example, the mn stacks might not all be identical. To stage them, there might be n copies having one kind of gate, another n having a different kind of gate, and so on. If the last instruction of each of the n stacks from collection i has an unlock for a stack from collection of $i+1$, we can simulate an iteration variable.

Alternatively, we can use a separate *variable stack* which is operated by unique strands.

For example, consider the situation in which the user would like to operate a walker, but progress the walker forward only one step per loop instance, releasing some set of instructions after each step. One could construct a stack consisting of the standard walker-driver instructions, but instead of being fueled by the general *tick* and *tock* strands, it will be fueled by a different pair of strands. We might refer to these strands as $call_0$ and $call_1$. These will be used to deploy the *tape* and *untape* strands (which must be called twice). Thus, the primary loop stack will be:

$$
\begin{aligned}
&gate_\mu/ \\
&\quad tick/ \\
&\quad tock/instruction_1 \\
&\quad tick/ \\
&\quad tock/instruction_2 \\
&\quad tick/ \\
&\quad tock/instruction_3 \\
&\quad tick/ \\
&\quad tock/call_0 \\
&\quad tick/ \\
&\quad tock/call_1 \\
&\quad tick/ \\
&\quad tock/call_0 \\
&\quad tick/ \\
&\quad tock/call_1 \\
&\quad tick/ \\
&\quad tock/unlock_\mu
\end{aligned}
$$

And the secondary variable stack will look like:

$$
\begin{aligned}
&call_0/ \\
&call_1/tape_{ax} \\
&call_0/ \\
&call_1/untape_{cy} \\
&call_0/ \\
&call_1/tape_{by} \\
&call_0/ \\
&call_1/untape_{ax} \\
&call_0/ \\
&call_1/tape_{cx} \\
&call_0/ \\
&\text{continues…}
\end{aligned}
$$

continued...
$call_1/untape_{by}$
$call_0/$
$call_1/tape_{ay}$
$call_0/$
$call_1/untape_{cx}$
$call_0/$
$call_1/tape_{bx}$
$call_0/$
$call_1/untape_{ay}$
$call_0/$
$call_1/tape_{cy}$
$call_0/$
$call_1/untape_{bx}$

The strategy is to have the instruction in the loop "call" the variable by way of issuing a "pop" instruction specific to the variable stack which stores the values of each interaction.

4.2.6   CONSIDERATIONS

Instructions will interact with some DNA target in the course of executing the DNA program. One challenge is to prevent this interaction from happening too early before the relevant instruction is released. One mechanism may simply rely on physical distance, perhaps by attaching the stack and the target to different rigid substrates, with enough separation so as to avoid interaction as long as both remain attached. When tock releases an instruction strand from the stack, it will be free to travel though the solution to its target machine in order to do its work.

Another method may be to rely on extrinsic knotting and linking topology. Instructions may be attached by both endpoints to the stack, thus requiring two clock cycles to fully release an instruction. This would prevent the instruction segment from binding prematurely, because the helical structure of duplex DNA would prevent it from wrapping around its target complement.

Fortunately, utilizing hairpin instructions, rather than linear strands would provide the same safeguards, while only requiring a single end to be bound to the stack. A large class of machines can be constructed using only hairpin-type fuels[Yin, Choi, Calvert and Pierce, 2008]. That is the approach we take in our experimental section.

An additional challenge is to avoid race conditions. If tick and/or tock strands are not completely consumed in the time they have to stabilize, some copies of the stack may release instructions too soon. For example, suppose tick strands remain when tock strands are added. As usual, an instruction will be released by a stack after the tock is added. But then, as a result of the leftover tick strands, a tick may hybridize to the same stack. If there are also leftover tock strands, then an additional instruction may be released. We call such a stack a *runahead* stack.

The basic approach to solving this problem is to remove excess unreacted strands. This might be accomplished by binding the tick/tock strands to a more easily manipulated substrate, such as a protein which can then be flushed In this approach, the tick and tock strands could be engineered to include sequences recognized by the binding sites of some protein, which can then be flushed from solution.

An alternative possibility is to bind the complement of the clock strands to some substrate, perhaps even collectively to a single large origami. After pouring tick into solution, for example, and allowing for sufficient time to stabilize, the perfect complement to tick would be added. While the complement of tick does contain a segment that can hybridize to the open toehold of a stack, the longer match with tick should dominate and displace any such hybridizations. This approach also has the advantage that once a clock strand has hybridized to its perfect complement, the resulting component will be completely stable, having no toeholds with which to react with another component.

Perhaps the simplest approach is to coat iron balls with complements to tick strands and to put those iron balls into the solution when it is time to flush away excess ticks. Another set of iron balls coated with tocks could also be used to flush away tocks. It is an experimental question to determine the best way to flush away extra tick/tocks.

4.2.7   EXAMPLE APPLICATIONS

4.2.7.1 Signal Amplification

A simple application of such stacks is to use them for signal amplification. This application uses a gated stack whose instructions are all the same. The gate-unlock strand opens the stack, and the ticks and tocks cause more and more copies of the instructions to be produced. Several different types of gates and amplified instructions may be used. Because the same instruction is repeated for the entire lifetime of the stack, runahead stacks are acceptable.

4.2.7.2 A Stack Controlled Walker

In the case of a cycled walker, a repeating sequence of 12 strands is required to operate the machine. This can become cumbersome and makes automation, such as by robotic pipettes, a complex proposition. By utilizing a looping stack module, such a walker may be driven by the simple alternating addition of tick and tock. This has the cost of requiring twice as many *pour* operations because releasing one fuel strand requires the application of tick and tock. However, the fact that there are only two fuel types required to drive the walker greatly simplifies lab procedures, reduces the possibility of user error, and simplifies automation.

These instruction stacks may be either attached to the same substrate as the walker, providing a self-contained system, or attached to a different substrate as a unique "driver module." The driver module can determine the direction the walker takes along the track by reversing the order of the tape-untape strands required to move the walker's feet. Thus, one module can be constructed for the "forward" direction, and another module—identical but for the reversal of instructions on the stack—can be constructed for the "backward" direction. By adding or removing these modules, the direction of the walker can be controlled while nonetheless using the same general tick and tock strands.

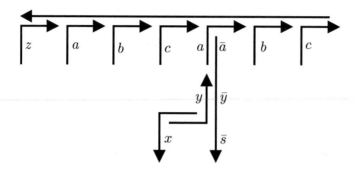

Figure 4.9: Operation of a clocked walker by way of a programmed stack.

Alternately, we might control a walker's directionality by using gated stacks. This is analogous to controlling the direction of the walker "at runtime" rather than "at compile time." Rather than using the looped module, we would use two types of gated stacks. One stack will have a gate for the "forward" direction, followed by a 12 instruction cycle. The other stack will have a gate for the "backward" direction and a similar 12 instruction cycle for moving the walker backwards. Initially, one of the two gate-unlock strands will be introduced into solution, and thereby unlock the appropriate stack. After 12 clock cycles, the desired direction will have to be chosen again by introducing another gate-unlock strand signifying the direction.

This idea can be extended to that of precisely controlling a walker on a two-dimensional grid. This requires two types of tracks: a "north-south" track and an "east-west" track. These two sets of tracks will intersect at every 6 footholds—12 instruction steps—in order to form a grid. By now utilizing 4 types of gated stacks, one for each of the cardinal directions, a clocked DNA computer can control a walker's movement about such a grid precisely. At each grid intersection, one of the 4 cardinal directions can be chosen by introducing the appropriate unlock strand.

4.3 FUTURE THEORETICAL WORK

Here is a summary of what we have accomplished so far:

1. A simple language to describe strands, their hybridization, ligation, strand loops, and so on.

2. A language describing the main transitions that are possible (e.g., hybridize, displace, exchange).

3. A method to combine elementary transitions with known motifs to support "stacks," each of which contains a sequence of transition instructions.

4. A further method based on gates on stacks to support conditionals, program looping, and subroutines.

With these steps, we have shown how to store programs inside DNA. The execution of these programs will require only clock ticks and tocks and some means to flush them. That is the good news. But it's just a beginning.

The primary benefit of the stored program computing concept is that instructions, data, and a problem input can all lie inside the computer. We have already seen how instructions can lie in a DNA soup as instruction stacks. Input DNA can simply be poured in. We have not yet discussed data structures. Data structures that store information basically have the structure of a key-value pair. Such structures support the query "What is the value associated with key k?" We can construct a DNA key-value data structure as follows: each key is implemented as a gate, and an associated value is implemented as a stack of strands that will be released with ticks and tocks. There are many variant ways to do this.

This framework implies that programming can begin. While the system we offer has instructions, conditionals, loops, and elementary subroutines, vastly different programming paradigms might be possible. We revisit this question after validating our ability to build clocked DNA stacks.

CHAPTER 5

Experiment: Storing Clocked Programs in DNA

5.1 CONSTRUCTING STORED INSTRUCTION STACKS

5.1.1 INTRODUCTION

In order to demonstrate that instructions can be stored within DNA[1] later to be released with clock strands, we created instruction stacks of various lengths. Constructing stacks is very similar to deconstructing them in that it involves pouring in alternating strands. For construction, however, we would start with the "base" strand and alternately pour anti-tick and instruction strands. This method, though time-consuming, proved to be the most reliable for making stacks of a determined length and, unlike many one-step methods, would allow for the construction of color-changing stacks.

Construction time is not a major concern because stacks can be prepared far in advance of their use and stored stably at low temperatures. In addition, construction (program compilation) can easily be automated, which would make it even more convenient.

Our findings indicate that this method of construction should be a viable method for making a collection of stacks all of the same approximate length. We have found that stoichiometric balancing of input strands can curb runahead chain reactions. Such chain reactions could occur when excess anti-tick and instruction strands continue to hybridize and create longer stacks than intended. As we have discussed in Chapter 4, stochiometric balancing (perhaps in coordination with a flushing mechanism) is also useful for the deconstruction step, again to counter runahead stacks. Each construction/deconstruction step requires slow cooling steps for accurate hybridization of added strands, but, again, construction time is not a major issue.

5.1.2 OVERVIEW OF METHODS

We first tried a one-step method, which combined base strands, instructions strands, and anti-tick strands in a 1:4:4 ratio to see if we could create stacks consisting of approximately four instructions. In this approach, we combined all the strands at 40°C to disrupt any unwanted base pairing. These pairings usually occur over small ranges (2-3 nucleotides) between strands during storage at low temperatures, when it becomes favorable for any sort of interaction to form between strands. We

---

[1] Aidan Daly carried out the work in this chapter in the summer of 2010 and wrote most of this description.

then cooled the combined strands slowly, as slow cooling favors correct hybridization between the longest matching sequences.

We ran the resulting construct on a native gel (one that allows DNA to exist in a double stranded state, thus not disrupting our constructs as would be the case for a denaturing gel). A gel permits us to determine the length of single and double strands. If there is just one size, there will be a single bar corresponding to that size. When there are many sizes, there will be a smear, where each part corresponds to one of the sizes.

The results (see Figure 5.1) showed a range of sizes, indicating that this one-pot method yielded many chain reaction constructions.

For this experiment, we used a small (8 bp), single-stranded base strand. The anti-tick strand was a 16bp single strand of DNA, and the instruction was a 56bp hairpin. Each of these components was run in an independent lane on the gel as a reference for size. The base strand was too small to appear on the gel and would dissociate from the stack fairly easily due to its small region of complementarity. The stack appeared as a large smear indicating the presence of many differently-sized constructs, containing anywhere between one and four instructions.

The smearing suggested that a sequential construction technique would be better. Sequential construction entails alternately pouring and hybridizing balanced amounts of anti-tick and instruction strands in order to control the length of the stack.

Because a sequential approach requires multiple rounds of heating and slow cooling (one for each strand to hybridize), we had to design a larger base strand. The single-stranded 8 base pair fragment would not complement a large range, and it would denature (i.e. the two complementary strands would split apart) from the bottom of the stack at relatively low temperatures. Thus, we designed a new base as a composite of two strands – SB (small base) and LB (large base) – with an overhang that bound to the first instruction of the stack. This increased size also allowed us to visualize the base on a gel.

Because we were performing a sequential construction, we developed a naming system to categorize all intermediates we would hope to observe on the gel (see Figure 5.2).

The sequential addition method consisted of the following protocol: heat equal molar amounts of the two base strands to a high temperature then cool slowly to allow hybridization. For each subsequent step, heat the current intermediate to a temperature high enough to disrupt unwanted base pairings, but not high enough to denature the construct. For example, when forming intermediate **C**, we combined an equal molar amount of anti-tick and **B** in a vessel and heated to just below the temperature when the top instruction strand of **B** would dissociate. These temperatures were determined by the length of complementarity between the strands, which, in this case, was 8 base pairs, as well as the GC content of the strands. The resulting mixture would then be slowly cooled to allow desired hybridization.

Following this method, we constructed **A** through **C** and ran the results on a native gel (see Figure 5.3.).

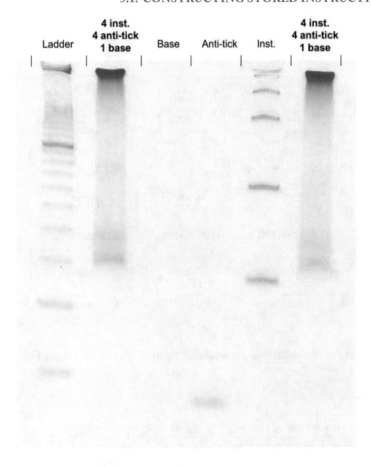

Figure 5.1: In this one step construction method, the stack appears as a large smear in both of the lanes we ran it in, indicating the presence of many differently-sized constructs, containing anywhere between one and four instructions. The one step construction method is therefore insufficient if we want all stacks to be of the same length.

The anti-tick, SB, LB, and construct **A** lanes all show clear bands, but once again, the lane containing the instruction strand shows multiple discrete bands. This is due to the formation of aggregate hairpins – multiple hairpins bound together. This occurs when the hairpin stem first melts, then binds to the melted stem domain of another hairpin, forming a double instruction, as illustrated below (see Figure 5.4). This can occur multiple times, and on the gel we see aggregates of up to six instructions forming in appreciable amounts.

For this reason, we see multiple bands in the lane consisting of construct **B** in Figure 5.5. We know this is not due to any chain reaction because no anti-tick had been added to that vessel. On

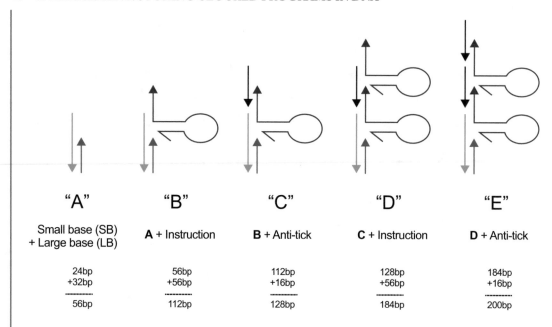

Figure 5.2: The results of a sequential construction method: precise selection of lengths.

the other hand, when we look at the band in the **B** lane representing only one instruction added (the band just above the one representing slight excess of intermediate **A**), we see there is a band just above it in the lane containing **C**, indicating that the hybridization of a stoichiometric amount of anti-tick would not have caused a chain reaction in any case.

In order to confirm this, we decided to perform the experiment with an additional heat shock step to separate aggregate hairpins. Before we added hairpins to any reaction vessel, we would heat the hairpin strands to a high temperature. This would open all hairpin stems and thereby separate all aggregate hairpins. Then, we would quickly cool the strands on dry ice. This quick cooling would favor intramolecular hybridization over intermolecular hybridization, ensuring that separate hairpins would not bind together. We then hybridized these instructions to the growing constructs with favorable results (see Figure 5.5).

The gel shows little aggregation of instructions. In solution, there should be even less aggregation, because the instruction strands used in the hybridizing reactions were used immediately whereas the ones run in the gel had rested some time since the heat shock. The single clear band in the lane for **B** indicates the incorporation of a single instruction onto the base.

The smearing observed around the band in lane **C** indicates some degree of chain reaction addition following the hybridization of anti-tick strands. However, the smearing is minimal and the

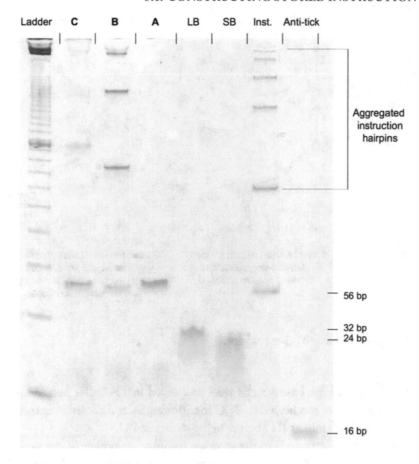

Figure 5.3: Results of the first attempt at a sequential addition method on a native gel – less of a smear but still not a single band.

band density is clearly focused slightly above the band for **B**, indicating that the incorporation of a single 16bp anti-tick strand is the major product.

5.1.3   PROTOCOL DETAILS

Strands were designed by hand and ordered from Integrated DNA Technologies. Strands were additionally purified by running on a denaturing polyacrylamide gel in a normal TBE buffer at 55°C, followed by standard n-butanol extraction.

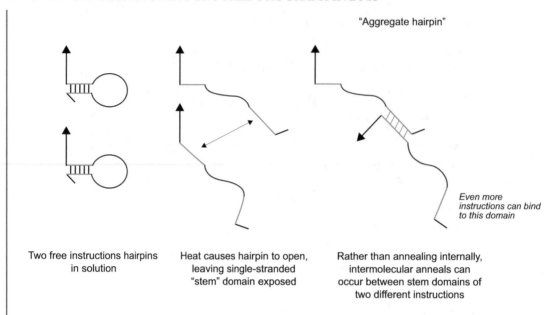

"Aggregate hairpin"

Two free instructions hairpins in solution

Heat causes hairpin to open, leaving single-stranded "stem" domain exposed

Rather than annealing internally, intermolecular anneals can occur between stem domains of two different instructions

Even more instructions can bind to this domain

Figure 5.4:   Hairpins can bind to other hairpins.

The small base and large base strands were suspended in TE buffer and combined in equal molar amounts. The mixture was heated to 95°C for 20 minutes and cooled incrementally (20 min at 65°C, 20 min at 37°C, 20 min at RT) before being stored at 4°C.

When adding anti-tick strands to the growing complex, a number of strands approximately equal to the number of moles of complex were measured out and the strands were dried using an evaporator. The complex (in solution) was added to the dried strands and resuspended using a vortex.

Instruction strands, prior to addition and hybridizing to the growing complex, were heated at 95°C for approximately 5 minutes, and then they were rapidly cooled on dry ice until addition to disrupt any aggregate hairpins. The instructions strands were incompletely dried, as complete drying would allow for re-formation of hairpin aggregates upon resuspension. The growing complex was added to the vial containing single instruction strands suspended in minimal 1x TE buffer and vortexed.

Hybridization for both anti-tick and instruction strands was accomplished by letting the strands sit at room temperature for 30 minutes, then storing at 14°C for at least 30 minutes.

At each hybridization step, a small aliquot of the intermediate product was removed and stored to be run on the native polyacrylamide gel alongside the final product. These changes in molar quantities of complexes were taken into account during subsequent strand additions.

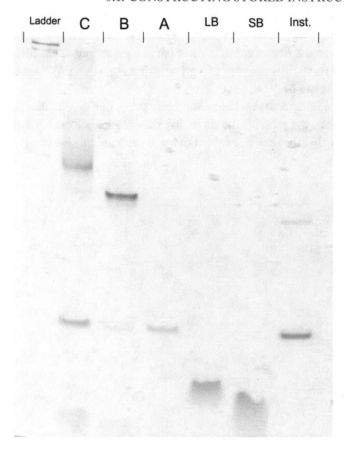

Figure 5.5: Hot-and-cold treatment eliminate hairpin aggregation, yielding a clear dominant band in lane **B**.

5.1.4   EXPERIMENTAL CONCLUSIONS/FUTURE DIRECTIONS

We have demonstrated that constructing stacks of a specific length (and therefore clocked DNA programs) can be achieved by a sequential addition process. We have also demonstrated that through simple stoichiometry, we are able to limit the chain reactions that occur due to excess instructions and anti-tick strands during construction. This may also have implications for stack deconstruction, where we seek to curb similar chain reactions with tick and tock strands. We have also demonstrated the usefulness of heat shock treatment of hairpin strands to ensure that aggregates are not incorporated into the stacks.

We have yet to determine how large we can make these stacks and how reliably we can control chain reactions at larger sizes. As they grow, native polyacrylamide gels may not be an informative enough means of visualization, as they are not capable of resolving small differences in size beyond a certain point. Fortunately, this is not a major concern because, for many applications, fine-grained accuracy is not important. For example, in most parallel processing applications, it will not matter if one stack releases more than one copy of a given instruction strand. We must also determine how large the gated stacks can be.

For the sake of simplicity, we may want to try using single-stranded instructions instead of hairpins. Nevertheless, the experiments show that the principle of storing instructions within the DNA in such a way that they can be released through clock strands is entirely realizable.

CHAPTER 6

A Clocked DNA Programming Language

6.1 EXISTING PROGRAMMING LANGUAGES MAY NOT WORK

In the first 10 years of their existence, automobiles resembled horsedrawn carriages – minus the horse hitches. This may have comforted early customers who enjoyed the familiarity of the vehicle, but it made no sense once the speed capabilities of the new technology became apparent.

While we have emphasized the familiarity of the clocked stored program concept in the preceding chapters, we do not mean to imply that programming languages for electronic computers should be used to control DNA computers. Let's first look at what is different about DNA computing.

- First, there is the question of scale. A large parallel electronic computer may have 1,000 processors or maybe even 10,000, but a clocked DNA computer can have billions of instruction stacks.

- Second, each step in an electronic computer takes under a billionth of a second to execute, whereas the billions of instruction stacks may take several minutes to stabilize after each tick and tock is released. So, even if the total raw computing power (instructions emitted per unit of time) of a DNA computer may be comparable to that of an electronic one, we must make every parallel instruction count.

- Third, the instruction stacks simply release instructions into solution. They keep no local state (there is nothing analogous to local variables or registers). All changes occur on the global state and on the stacks themselves. This means that a "call" to a gated instruction stack using an unlock strand does not pass parameters.

- Fourth, as we discussed in Chapter 4, it is possible to pour in a vial of tick strands, but they may not all be consumed. When tock strands are poured in next, some stacks may release several instructions while others release none. So, clocking can be imprecise.

The strategy suggested in Chapter 4 was to administer strands in proper stoichiometry and then wait long enough. Other strategies that may be more precise are possible. For example, a vial may be administered that contains more tick strands than instruction stacks, and then some flushing mechanism (such as coated iron balls) may remove excess ticks. A programmer should be able to specify whether such a *flushing* mechanism is necessary.

• Fifth, a stack with a gate will not execute unless it receives an unlock strand. It is conceivable, however, that some gated stacks will receive unlocks at a different time than other ones. A programmer should be able to specify whether all stacks sharing the same gate should begin at the same or whether this common start time is unnecessary.

6.2   PARAGAT: A MINIMALIST LANGUAGE FOR A CLOCKED PARALLEL DNA COMPUTER

The language *ParaGAT* consists of instructions (which are strands of DNA), gates (which are other strands), and synchronization primitives. The basic synchronization primitive is **flush**, which entails ensuring that there are no clock strands (ticks or tocks) in the system. Flushing is the primitive used to ensure that instructions are executed one at a time. The notion that instructions must be executed synchronously occurs so frequently, however, that *ParaGAT* also includes the keywords **sync** (meaning the ith instruction in every stack should execute before any stack executes instruction $i + 1$) and **async**.

The keyword **async** means that the programmer allows different instances of some stack to execute different instruction numbers concurrently. Operationally, this means that the system can flood with tick and tock strands to allow runaway stacks. (It might be possible to use some version of the asynchronous Qian stacks [Qian, Soloveichik and Winfree, 2011]. Qian stacks once unlocked (if necessary) would have a sequence of instructions each released by a complement of an instruction and a toehold. The complements would be pre-loaded in the system. It is an engineering question whether Qian stacks or the hairpin style stacks proposed here would be a better implementation of asynchronous stacks.)

The keyword **async** also implies that the stack construction step is allowed to construct stacks of different sizes provided the stacks differ only in how many times a given instruction is repeated. For example, for a parallel async stack containing the nominal sequence of instruction strands X Y X Z could also have stack instances having the sequence X X Y Y X Z Z. This freedom may permit fast parallel construction methods like those that caused a smear in the last chapter.

The notation is very simple. For example, the following describes a block of ungated synchronous parallel stacks. Each payload instruction is just a DNA strand.

```
begin sync
   payloadInstruction 1
   payloadInstruction 2
   payloadInstruction 3
   payloadInstruction 4
end sync
```

Next, *ParaGAT* should have a way to specify a block of instructions representing a collection of gated stacks. The specification should include the gate condition required to unlock each stack. Here the condition will be a strand or strands that together will unlock the stack. (It is easy to build a stack that requires the Boolean *and* of a set of strands to open. Each corresponding lock is just another gate.)

A block of asynchronous stacks that will be unlocked if the strand g appears will take the form:

```
gate g
begin async
  payloadInstruction 1
  payloadInstruction 2
  payloadInstruction 3
  payloadInstruction 4
end async
```

This same gating idea is also used for the special instruction **lookup**. The lookup instruction takes a strand as an argument. The strand is the key that unlocks the key-value stack. Lookup returns the value in that stack.

As we will see below, it is sometimes necessary to ensure that all instances of the gated stacks begin executing at the same time (even when the gated stacks themselves will execute themselves asynchronously). One way to ensure that is to pour in part of the gate as needed. Admittedly, this means that, in addition to tick and tock, other strands may be poured into the DNA solution. These strands will often be used as higher level clock strands, e.g., a strand to indicate the computation is in phase k. As long as this is done sparingly, it could be very useful. We call this primitive **external**; it takes a strand as an argument.

When the programmer wants all stacks having this gate to begin executing at the same time, we make the gate external and introduce it only after we can be sure the previous stacks have completed.

```
external gate g
begin async
  payloadInstruction 1
  payloadInstruction 2
  payloadInstruction 3
  payloadInstruction 4
end async
```

A program is written as a hopefully small collection of gated and ungated blocks. This "dataflow" style of programming, in which strands are released when their pre-conditioning gates are satisfied, models the DNA in solution nicely.

6.2.1 CONCEPTUALIZING COMPUTATIONS IN PARAGAT

One way to view synchronous instruction stacks in *ParaGAT* is as a sequence of instructions having a strict linear dependency. Conversely, asynchronous instruction stacks have no inter-dependencies.

Therefore, a more convenient way to model the computation is as a graph where each vertex is an instruction (a single DNA strand ultimately), a possibly non-empty gate (one or more single DNA strands), and possible dependencies. Edges denote dependencies. So $x \rightarrow y$ means that instruction x must precede instruction y.

This graph representation compiles to ParaGAT quite simply. Instructions having common gates are put into the same gated stack. Instructions having no gates are put into a single ungated stack. The stack is synchronous if some of the instructions of a stack have dependencies on other instructions in the stack. The stack is asynchronous otherwise. A gate g is external if all instructions either directly or transitively precede or follow g.

6.3   AN EXAMPLE APPLICATION IN PARAGAT

As a practical example of our approach, consider a detector for several harmful single stranded pieces of DNA. Let's say that we want to indicate which strand is present by using quenched fluorescent markers. So, a DNA instruction in a stack will not fluoresce but will do so once released. Now, suppose we have two colors (red and blue) to choose from, but we want to distinguish between four different strands, which we will call $s1$, $s2$, $s3$, and $s4$.

We arrange our stacks so that $s1$ and $s2$ will cause red to show in a first time period given the external gate "phase1." Then $s1$ will cause red to show in the second (following) time period, but $s2$ will cause blue to show in the second time period. Similarly, $s3$ and $s4$ will cause blue to show in a first time period. Then $s3$ will cause red to show in the second time period, but $s4$ will cause blue to show in the second time period.

Here is the block for the first time period for $s1$ and $s2$. (We have not yet figured out how to construct a gate that embodies a boolean *or*. So, we may in fact need different stacks for the *or* condition.)

```
gate (s1 or s2)
external gate phase1
begin async
  red fluorescent marker instruction
  red fluorescent marker instruction
  red fluorescent marker instruction
  red fluorescent marker instruction
end async
```

We treat $s3$ and $s4$ similarly. They cause a blue fluorescence in the first time period.

```
gate (s3 or s4)
external gate phase1
begin async
  blue fluorescent marker instruction
  blue fluorescent marker instruction
```

```
      blue fluorescent marker instruction
      blue fluorescent marker instruction
   end async
```

For the second time period, we use an external gate to ensure that the second time period stacks don't start emitting their markers while the first time period stacks are emitting their strands. That is, we might flush out the fluorescence from the first time period as well as the phase1 strands and then introduce the phase2 strand externally. Either $s1$ or $s3$ cause red fluorescence in the second time period.

```
   flush away fluorescence and phase1 strands
   gate (s1 or s3)
   external gate phase2
   begin async
      red fluorescent marker instruction
      red fluorescent marker instruction
      red fluorescent marker instruction
      red fluorescent marker instruction
   end async
```

Either $s2$ or $s4$ cause blue fluorescence in the second time period.

```
   flush away fluorescence and phase1 strands
   gate (s2 or s4)
   external gate phase2
   begin async
      blue fluorescent marker instruction
      blue fluorescent marker instruction
      blue fluorescent marker instruction
      blue fluorescent marker instruction
   end async
```

6.4   FUTURE WORK

ParaGAT is essentially a machine language. It consists of strands to be put on stacks, strands to act as gates, and primitives for flushing and external unlocking. Skilled language designers will be able to build higher level constructs on top of these primitives. The challenge is to construct a language that is high level and easy to write, but that also compiles to time- and material-efficient DNA programs. At this point, we see this as a wide open problem.

Bibliography

Leonard Adleman. Molecular computation of solutions to combinatorial problems. *Science* 266(1994) 1021-1024. DOI: 10.1126/science.7973651 2

Martyn Amos. *Genesis machines: the new science of biocomputing*. Atlantic Books, November 16, 2006 ISBN 1843542242. 6

Jonathan Bath and Andrew J. Turberfield. DNA nanomachines. *Nature Nanotechnology*, 2:275–284, May 2007. DOI: 10.1038/nnano.2007.104 6

Y. Benenson, T. Paz-Elizur, R. Adar, E. Keinan, Z. Livneh, and E. Shapiro. Programmable and autonomous computing machine made of biomolecules. *Nature* 414 (2001), 430-443. DOI: 10.1038/35106533 5

C. H. Bennett. Logical Reversibility of Computation. *IBM Journal of Research and Development* 17 (1973) 525-532. DOI: 10.1147/rd.176.0525 2

Banani Chakraborty, Ruojie Sha, and Nadrian C. Seeman. A DNA-based nanomechanical device with three robust states. *PNAS*, 105(45):17245–17249, November 2008. DOI: 10.1073/pnas.0707681105 10

Matthew Cook, Paul W.K. Rothemund, and Erik Winfree. Self-assembled circuit patterns. *Lecture Notes in Computer Science*, 2943:91–107, 2004. DOI: 10.1007/978-3-540-24628-2_11 10

Shawn M. Douglas, Adam H. Marblestone, Surat Teerapittayanon, Alejandro Vazquez, George M. Church, and William M. Shih. Rapid prototyping of 3D DNA-origami shapes with caDNAno. *Nucl. Acids Res.*, 37(15):5001–5006, 2009. DOI: 10.1093/nar/gkp436 10

Christofer Hedbrandh. Using the A^* algorithm to build DNA nanostructures. Master's thesis, Chalmers University of Technology, February 2010. 22

M. Ogihara and A. Ray. Simulating Boolean circuits on DNA computers. *Algorithmica* 25 (1999), 239-250. DOI: 10.1007/PL00008276 5

Lulu Qian, David Soloveichik, and Erik Winfree. Efficient Turing-Universal Computation with DNA Polymers. *Lecture Notes in Computer Science 6518* pp. 123-140, 2011. DOI: 10.1007/978-3-642-18305-8_12 5, 52

Paul W. K. Rothemund. Folding DNA to create nanoscale shapes and patterns. *Nature*, 440(7082):297–302, March 2006.

10, 30

Paul W.K. Rothemund. A DNA and restriction enzyme implementation of Turing machines. In *DIMACS Series in Discrete Mathematics and Theoretical Computer Science*, volume 27, pages 75–119, 1996. 5, 10

Nadrian C. Seeman. Nucleic-Acid Junctions and Lattices. *Journal of Theoretical Biology* 99, no. 2 (1982) 237-247. DOI: 10.1016/0022-5193(82)90002-9 2

Dennis Shasha and Cathy Lazere. *Natural Computing: DNA, Quantum Bits, and the future of smart machines.* W. W. Norton, 2010 ISBN-10: 0393336832. 1

William B. Sherman and Nadrian C. Seeman. A precisely controlled DNA biped walking device. *Nano Letters*, 4(7):1203–1207, 2004. DOI: 10.1021/nl049527q 28

Jong-Shik Shin and Niles A. Pierce. A synthetic DNA walker for molecular transport. *Journal of the American Chemical Society*, 126(35):10834–10835, 2004. DOI: 10.1021/ja047543j 10, 28

Soo-Yong Shin, Byoung-Tak Zhang, and Sung-Soo Jun. Solving traveling salesman problems using molecular programming. *Proceedings of the Congress on Evolutionary Computation*, 2:994, 1999. DOI: 10.1109/CEC.1999.782531 10

Suvir Venkataraman, Robert M. Dirks, Paul W. K. Rothemund, Erik Winfree, and Niles A. Pierce. An autonomous polymerization motor powered by DNA hybridization. *Nature Nanotechnology*, 2:490–494, August 2007. DOI: 10.1038/nnano.2007.225 28

Erik Winfree. On the Computational Power of DNA Annealing and Ligation. DNA Based Computers, pp. 199-212, 1996. DOI: 10.1016/S0166-218X(96)00058-3 10

Erik Winfree, Xiaoping Yang, and Nadrian C. Seeman. Universal Computation via Self-assembly of DNA: Some Theory and Experiments. DNA Based Computers II, pp. 191-213, 1998. 10

Peng Yin, Harry M. T. Choi, Colby R. Calvert, and Niles A. Pierce. Programming biomolecular self-assembly pathways. *Nature*, 451:318–322, January 2008. DOI: 10.1038/nature06451 38

B. Yurke, A.J. Turberfield, A.P. Mills, F.C. Simmel, and J.L. Neumann. A DNA-fuelled molecular machine made of DNA. *Nature* 406:605-609, 2000. DOI: 10.1038/35020524 6, 27

David Yu Zhang and George Seelig. Dynamic DNA nanotechnology using strand-displacement reactions. *Nature Chemistry* 3 103-113, 2011. DOI: 10.1038/nchem.957 27

Y. Zhang and Nadrian C. Seeman. The Construction of a DNA Truncated Octahedron. *Journal of the American Chemical Society* 116 (1994): 1661-1669. DOI: 10.1021/ja00084a006 2

Authors' Biographies

JESSICA P. CHANG

Jessica P. Chang received her B.S. in Mathematics from the University of Texas at Austin in 2007 where she developed an interest in topology and computer science. After completing an independent study course in topology under Prof. Genevieve Walsh, she became interested in applying topology to real-world problems. As an undergraduate, she researched topological modeling of DNA under the direction of Prof. Isabel Darcy. During this time she co-authored a publication on using knot theory to model DNA-protein complexes. She then went on the obtain her M.S. in Mathematics from New York University in 2010. There she worked with Prof. Dennis Shasha applying her knowledge of knot theory and DNA to the present work.

DENNIS SHASHA

Dennis Shasha is a professor of computer science at the Courant Institute of New York University where he works with biologists on pattern discovery for microarrays, combinatorial design, network inference, and protein docking; with physicists, musicians, and financial people on algorithms for time series; and on database applications in untrusted environments. Other areas of interest include database tuning as well as tree and graph matching. He has written four scientific books about database tuning, biological pattern recognition, time series, and statistics. For fun, he has also written six books of puzzles about a mathematical detective, a biography of great computer scientists, and a book about the future of computing. He has co-authored over sixty journal papers, seventy conference papers, and fifteen patents. He has written the puzzle column for various publications including *Scientific American.*

Index